Deliverance

Hassan Rasheed

Deliverance

ISBN
978-1-365-79353-0

Printed in the United States of America
By
www.lulu.com

Table of Contents

Chapter   Page

Introduction

In 1972, *The Limits to Growth* study addressed how humans would adapt to the physical limitations of planet Earth. It forecast that from the year 2000 to 2050 the ongoing growth in the human ecological footprint would stop-either through catastrophic "overshoot and collapse" or a well-managed "peak and decline."

In this book we look at the root causes of our needs for a high rate of consumption, which is not foreign to other plants and animals. As a "successful" species with unique talents we have failed to recognize, even to this day, that we need to adopt a "big picture" approach of looking at Earth and its biota.

It is imperative that we slowdown from our daily routines and recognize ourselves as part of and not separate from this beautiful life giving planet. We need to focus on human well-being rather than on per capita income growth. We will be constrained into unfortunate situations in surprising ways if we do nothing.

So, how do we prepare for the years ahead? With heart, fact, and wisdom, I will guide you along a realistic path from the distant past into the future and discuss what readers can do to ensure a better life for themselves and their children during the increasing turmoil of the next 30 years.

Presently, we need to look at life, its origins and its modes-operandi.

Preamble

May the Cycle be Unbroken

In the beginning, 4 billion years ago, there was a dance on Earth to the laws of physics by rock, wind, and water powered by the rays of the sun. A person could say the Earth was alive with rain falling, rivers flowing, and landscapes transforming: all dancing in cycles. Yet to Mother Earth, it was not alive enough, and so new dance cycles came into being 3.8 billion years ago guided by genetic laws of the organic matter. As the number of organic dance cycles grew, they started to interlock, forming circular dancing chains that too interconnected: a kaleidoscope of colors and forms. This state of affairs transformed Earth's surface into stages of green for the dramas of the song of birds, the roar of bull elephants, and the discourse of tribal leaders to name just a few and far between.

The fruits of one dance cycle were given freely to other dance cycles that then recycled them back to Earth. Again, sunlight fueled these new dance cycles. Each new organic life form grew through the stages of thriving, expanding, acclimating, and maturing hand in hand with each other and with the finite Earth. Each thrived on the available resources with added efficiency, expanded its territory when it had the chance, adjusted its lifestyle by finding its ideal habitat, and matured in a way that added stability to its environment by teaching its children to dance to the circular rhythm of the Sun–Earth beat.

In a blink of an eye 10,000 years ago, dance cycles started to break. Chains unraveled, stages collapsed, and colors faded. "Why, oh why was this happening?" exclaimed Mother Earth as she cried and searched for the answer. Finally, she discovered a species that was stuck in the expansion stage and could not get to the stage of maturation to continue the cyclical dance. This species developed endless self-gratification to the point it could not see how it was detrimentally affecting its dance and other dance cycles and breaking up the fabric of life on Earth. In many instances, the dance of this species stopped the dance of others who then vanished.

When Mother Earth questioned this species on their behavior, it responded, "We have the right to be free and to promote freedom around the world. No one is going to tell us what to do." It did not understand that it needed to continue growth through all dance stages. Maturation added stability to Earth for its children and the children of others. When questioned on that point it only came back with, "I can manage my own dance and the dance of others on Earth just fine through the use science and technology. I can do it. Just look at my achievements. I have conquered the Earth and even the moon!"

The Earth was powerless to change this immature species' mind. After all, they originated from her, and she knew she could not change herself.

Abiogenesis

Informally, the origin of life or abiogenesis is the natural process by which life arises from nonliving matter, such as simple organic or inorganic compounds. It is thought to have occurred on Earth between 3.8 and 4.1 billion years ago. Abiogenesis is studied through a combination of laboratory experiments and extrapolations from the characteristics of modern organisms, and aims to determine how pre-life chemical reactions gave rise to life on Earth.

In simple terms, life is a chemical reaction requiring energy. With a fixed amount of incoming energy from the Sun (approximately 3,459,000 calories per square yard) hitting the Earth's surface, lowering the amount of energy required for this chemical reaction to take place is what evolution is all about. If you ask how does lowering the amount of energy in a chemical reaction is achieved, the answer is in the availability of a catalyst. In most cases where a catalyst is required, reactions occur faster because they require less activation energy (Please refer to Appendix 6 for more information on catalysts and evolution).

In summary, evolution is self-replication of catalysts where a change or changes occur to the offspring that lead to more energy efficient catalysis.

Abiogenesis proposes that the first life-forms generated were very simple and through a gradual process became increasingly complex. In the 1920s British scientist J.B.S. Haldane and Russian biochemist Aleksandr Oparin independently set forth the belief that organic molecules could be formed from abiogenic materials in the presence of an external energy source such as ultraviolet radiation and that the primitive atmosphere had very low amounts of free oxygen, contained ammonia and water vapor, among other gases. Both also suspected that the first life-forms appeared in the warm, primitive ocean and were obtaining preformed nutrients from the compounds in existence on early Earth rather than generating food and nutrients from sunlight or inorganic materials.

Oparin believed that life developed from microscopic spontaneously formed spherical aggregates of lipid molecules that are held together by electrostatic forces and that may have been precursors of cells. Oparin's work confirmed that enzymes fundamental for the biochemical reactions of metabolism functioned

7

more efficiently when contained within membrane-bound spheres than when free in aqueous solutions. Haldane, unfamiliar with Oparin's work, believed that simple organic molecules formed first and in the presence of ultraviolet light became increasingly complex, ultimately forming cells.

In 1953 American chemists Harold C. Urey and Stanley Miller tested the Oparin-Haldane theory and successfully produced organic molecules from some of the inorganic components thought to have been present on prebiotic Earth by combining warm water with a mixture of four gases—water vapor, methane, ammonia, and molecular hydrogen. They then subjected this mixture to electrical discharges. One week later Miller and Urey found that simple organic molecules, including amino acids which are the building blocks of proteins, had formed under the simulated conditions of early Earth.

Modern abiogenesis hypotheses are based largely on the same principles as the Oparin-Haldane theory and the Miller-Urey experiment. There are, however, subtle differences between several models that have been set forth to explain the progression from abiogenic molecule to living organism. One explanation says that complex organic molecules first became self-replicating entities lacking metabolic functions. The other says that metabolizing protocells grew first then developed the ability to self-replicate.

Self-replication requires the ability to "remember" how to make another copy of one's self. This requires some sort of memory device. On Earth this molecule or molecules are made of nucleic acid as will be shown in the next chapter on the cell.

The Cell

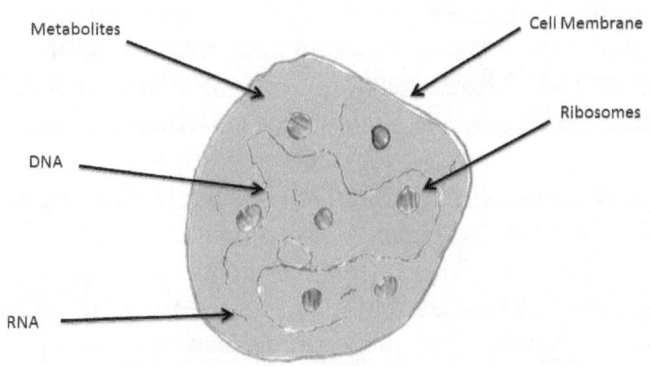

Mycoplasma genitalium

Mycoplasma genitalium, a parasitic micro catalyst bacterium (See Appendix 5 for terminology use) which lives in the primate bladder, waste disposal organs, genital, and respiratory tracts, is thought to be the smallest known catalytic organism capable of independent growth and reproduction. With a size of approximately 200 to 300 nm, *M. genitalium* is an ultra-micro-bacterium, smaller than other small bacteria.

M. genitalium is a far cry from a simple enzyme or catalyst. It has several structures that all work together to form a super catalyst. The cell membrane (also known as the plasma membrane or cytoplasmic membrane) is a biological membrane that separates the interior of all cells from the outside environment. The cell membrane is selectively permeable to ions and organic molecules controlling the movement of substances in and out of cells. The basic functions of the cell membrane are to protect the cell from its surroundings and to maintain an optimal internal environment for survival and reproduction.

The internal environment or plasma contains 5 basic structures. First, there are the metabolites or the inputs to the catalytic reactions that occur in the cell. Also present are the products of the catalytic reactions, namely proteins, which are in turn used to build larger structures such as ribosomes.

A ribosome is a particle consisting of ribonucleic acid (RNA) intermeshed with proteins, found in large numbers in the plasma. Its function is to catalyze transformation of metabolites into protein products. You may ask, "How does a ribosome make different proteins?" To answer this question, we need to visit the DNA found in the cell. DNA is deoxyribonucleic acid, which is a molecule or molecules found in the plasma that contain every formula on how to make cell proteins. The formula is in the form of a sequence of codes that are transcribed into an RNA sequence of codes that spell out how the ribosome is to make any particular cellular protein.

All functions of the cell require energy (see Appendix 2) and because the limited source of energy is the sun, it is only logical that those cells that are more efficient in the use of available energy will succeed and outpace less efficient cells when it comes to growth and reproduction and will eventually dominate the environment as was demonstrated in the section on catalysts. In fact, one can look at the cell of *M. genitalium* as a complex catalyst.

An observation can be made at this time that one trend in evolution is that of an increase in complexity of living organisms. For example, the sketch below shows a more complex cell that is thought to have evolved from *M. genitalium*-type cells.

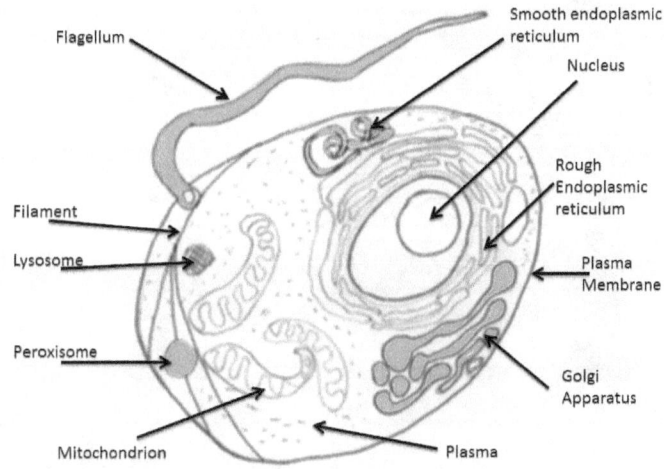

Flagellum

Smooth endoplasmic reticulum

Nucleus

Rough Endoplasmic reticulum

Filament

Lysosome

Plasma Membrane

Peroxisome

Golgi Apparatus

Mitochondrion

Plasma

Complex Single Celled Organism

Can we also say that more complex cells are more energy efficient than less complex ones as catalysts?

Another observation we can make looking at the complex cell above is that all its individual parts are cooperating in unison to form a living entity. Can we call this "working together" a form of cooperation?

Last but not least is the observation that in more complex cells, a nucleus forms that houses DNA, which contains information in the form of sequences of amino acids (basic building blocks of proteins) required to construct every protein and structure in the cell. Some people like to say that the nucleus with its DNA is the most important part of the cell. The truth of the matter is that all parts of the cell are as important as the nucleus. No one part of the cell can exist without the other parts.

Evolution

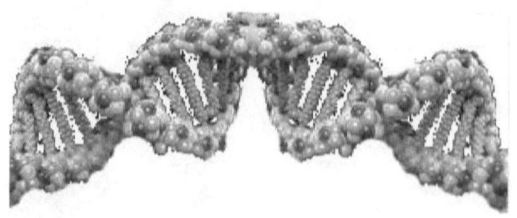

Evolution is the source of the different species that inhabit the various ecosystems around the globe. But how does evolution do that?

The picture above depicts a very small section of DNA. DNA is partially responsible for evolution. It contains the code that dictates such things as how an individual develops his or her physique and behaviors.

The environment is the other part of the equation responsible for evolution. It provides variation or change to DNA. This variation comes from γ-rays and toxins (and other chemicals) in the surrounding environment. The environment also provides the habitat in which species find shelter and food and are given the opportunity to compete for resources.

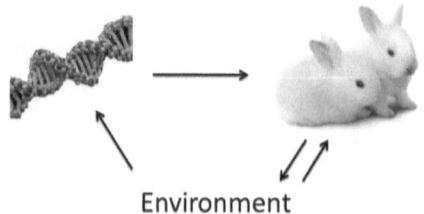

Environment

In the last diagram, which describes the vectors of evolution, the environment affects the genetic code, the genetic code affects the development of the rabbit, and the rabbits affect the environment. There are two opposing arrows between the environment and the two rabbits, indicating that the rabbits influence the environment because they dig holes and eat vegetation, while the environment influences the rabbit by providing the habitat.

If there is a change to the environment that alters the habitat of a species, the species may migrate away to a more suitable habitat, adapt to it, or die. This means the species will have to change physically or behaviorally to survive, and therefore there needs to be a change or changes to its DNA. Luckily, the environment provides the means for changing DNA.

A mutation is a random change in DNA. An organism's DNA affects how it looks, how it behaves, and its physiology. Thus, a change in an organism's DNA can cause changes in all aspects of its life. Mutations are essential to evolution; they are the raw material of genetic variation.

Natural selection is the process by which nature selects the organisms whose mutations make them better able to survive long enough to reproduce and thus pass on the same genes to their offspring.

Complexity

Evolution has produced some remarkably complex organisms and organism communities although the actual level of complexity is hard to define or measure accurately in biology, with properties such as DNA content, the number of cells, cell types, life cycle, associations with other organisms, and/or morphology all being used to assess an organism's or community's complexity.

Paramecium

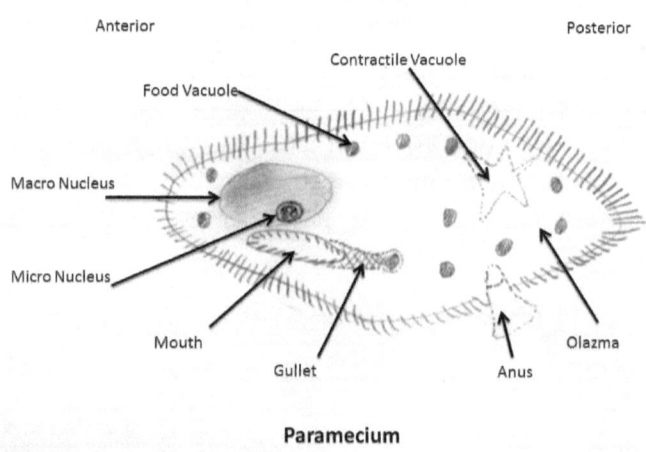

Paramecium

Paramecium is a single-celled micro catalytic organism with hairlike cilia covering its body that help it move around usually in fresh water although there are some species inhabit brackish or salty waters. To better capture their prey, they eject vacuoles or trichocysts filled with proteins. Trichocysts can also be deployed for self-defense by distracting enemies.

Paramecium cannot manufacture its own food and instead obtains its food and energy by taking in organic substances, usually plant or animal matter. Its common form of prey is bacteria. A single organism has the ability to eat 5,000 bacterial cells per day.

Paramecium is also known to feed on yeast, algae, and small protozoa. It captures its prey via a process whereby an individual engulfs a solid particle to form an internal compartment known as a phagosome.

Paramecium is capable of both sexual and asexual reproduction. Asexual reproduction is the most common and is accomplished by the organism dividing transversely. *Paramecium* can reproduce asexually two or three times a day. Normally, *Paramecium* reproduces sexually only under stressful conditions. This process occurs via conjugation, a process of gamete agglutination and fusion where two *Paramecium* cells join together, forming a conjugation bridge.

Cooperation

Cooperation in evolution is when two or more organisms of the same or different types or species work or act together for common benefits; a sort of a one for all and all for one kind of behavior. This process is the opposite of inter- or intragroup competition where individuals work against each other. The diversity of taxa that exhibit cooperation is quite large, ranging from single cells, to insects, birds, and mammals like African elephants.

An example is the slime mold amoeba. Slime mold is an informal name given to several kinds of unrelated eukaryotic micro catalytic organisms that can live freely as single cells, but aggregate together as a mezzo catalyst organism to form multicellular reproductive structures. Slime molds are formerly classified as fungi but are no longer considered a part of that kingdom. Although not related to one another, they are still sometimes grouped for convenience within the paraphyletic group referred to as kingdom Protista.

More than 900 species of slime mold occur all over the world. Most slime molds are smaller than a few centimeters, but some species may reach a size of up to several square centimeters and mass of up to 30 grams.

They feed on microorganisms that live in any type of dead plant material. They contribute to the decomposition of dead vegetation and feed on bacteria and fungi (such as yeast). For this reason, slime molds are usually found in soil, lawns, and on the forest floor, commonly on deciduous logs. In tropical areas, they are also common on inflorescences, fruits, and in aerial structures (e.g., in the canopy of trees). In urban areas, they are found on mulch or even in the leaf mold in gutters and also grow in air conditioners, especially when the drain is blocked.

This organism has three morphological forms associated with its life cycle. First, it exists as a single-celled, independent organism feeding on bacteria in the soil, thus obtaining the energy to multiply. Second, when food runs out, a large group of slime mold cells aggregate (through a process called streaming) and form a multicellular slug that moves in a certain direction. Third, the slug forms a sedentary long stalk capped with a fruiting body. The individuals in the fruiting body become spores that are carried away by wind or other organisms. As the spores land in a new area, individuals emerge and start to feed and multiply.

The aggregation process of slime mold involves the release of a chemical into the environment called cyclic adenosine monophosphate (AMP) by one amoeba. This chemical signals other amoebas to "stream" or move towards it to build the slug.

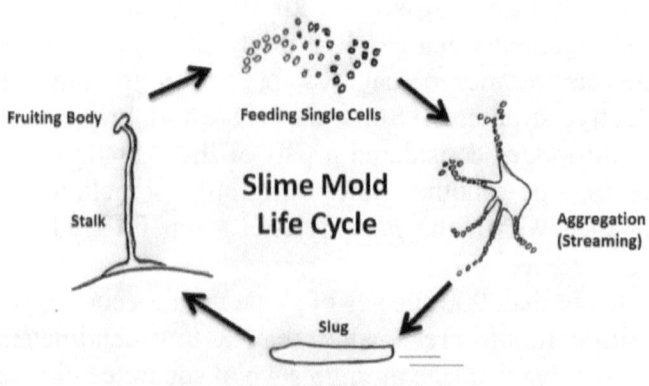

Three alternate morphologies and behaviors of the slime mold, namely, feeding, aggregation to form the slug and then stalk, and the fruiting body.

The slug looks pretty much like a human finger without the nail. At this stage, it is a multicellular organism. It has anterior and posterior ends and moves in one direction. To achieve this, the individuals making up the slug must communicate with one another on at least two levels. The first is what position in the slug each individual will take. The second is how to coordinate the rhythmic actions of each individual cell to cause the slug to move from one place to the next.

The fruiting body's support stalk is made up of 100,000 individuals or more. The stalk lifts up the fruiting body high, and the latter contains the spores. The support stalk individuals give up the ability to disperse or have progeny, yielding this opportunity to the fruiting body spores alone. They, the stalk cells, have demonstrated the extreme cooperative behavior of altruism, which is the sacrifice of oneself for the benefit of others because the stalk eventually weathers away and dies.

Here, cooperation serves to help the slime mold genome find greener pastures of energy.

Earthworms

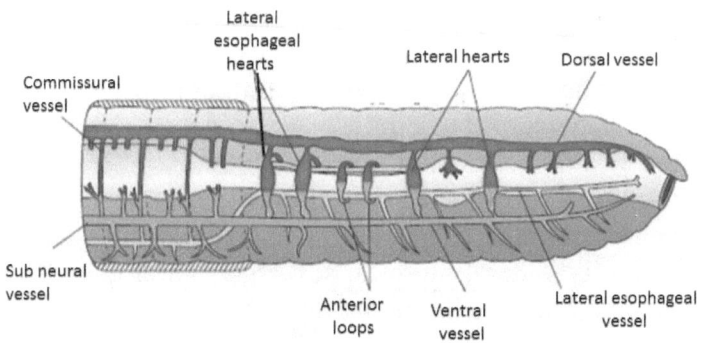

Earthworms are multicellular mezzo catalytic organisms that are more complex than the slug of the slime mold. Its cells cooperate and are differentiated (specialized) and organized into organs, each of which has a specific function. For example, it has a circulatory system to transport oxygen and nutrients to where they are needed. It also has a mouth and a gut for ingesting and processing its food.

Earthworms are reddish brown terrestrial invertebrates that inhabit the upper layer of moist soil. During daytime, they live in burrows made by boring and swallowing the soil. In gardens, they can be traced by their fecal deposits known as worm castings. Earthworms have a long cylindrical body. The body is subdivided into more than a hundred short segments that are similar (metameres ~100–120 in number). The dorsal surface of the body is marked by a dark median mid-dorsal line (dorsal blood vessel) along the longitudinal axis of the body. They can also move about by means of their muscles that manipulate the segments.

The ventral surface is distinguished by the presence of genital openings (pores). The anterior end consists of the mouth and the prostomium, a lobe which serves as a covering for the mouth and as a wedge to force open cracks in the soil into which the earthworm may crawl. The prostomium has a sensory function.

The body wall of the earthworm is covered externally by a thin non-cellular cuticle below which the epidermis is located, followed by two muscle layers (circular and longitudinal) and an innermost coelomic epithelium. The epidermis is composed of a single layer of columnar epithelial cells, which contain secretory gland cells. The alimentary canal is a straight tube and runs from the first to last segment of the body.

With regard to the earthworm, the permanent cooperation of its cells, their specialization, and formation of organs such as the gut, mouth, nervous system, muscles, sex organs, and glands apparently contribute to its success in finding the energy-providing food it needs to survive and reproduce.

Hassan Rasheed

Symbiosis

Symbiosis is a close association between two or more organisms of different species that often but not necessarily benefit each other. The association of algae and fungi in lichens and of bacteria living in the intestines or on the skin of animals are forms of *symbiosis*.

Leafcutter ants constitute 41 mezzo catalytic species of ant across genera *Atta* and *Acromyrmex* and represent the most socially complex division of the fungus-farming ants, which encompasses some 200 species of the family Formicidae called the attine ants, sometimes referred to as the Attini tribe. Indeed, their unique symbiotic relationship with a fungus has been compared to the agricultural practices of humans!

Colonies consist of up to 8 million individuals, with significant polymorphisms based on specialized roles, or castes, within the nest. Their underground nests can range from 30 to 600 meters in diameter, with smaller rounds radiating from a central mound that can itself be 30 meters in diameter. All leafcutter species are endemic to Southern and Central America and parts of the Southern United States and are responsible for consumption of 20% of the fresh leaf bio-matter of the Noosphere.

Leafcutter ants are not leaf-eater ants. Their herbivorous efforts go into feeding fungal cultivars of the Lepiotaceae family scrupulously maintained within the nest. Fungal hyphae secrete digestive enzymes into the fresh leaf substrate, transforming cellulose in the leaves into an accessible form of ant food. Swelling with sugars and protein, the hyphal tips, called "staphylae" or "gongylidia," are then consumed by the ant and their larvae.

In this case of symbiosis, the ants feed the fungus the energy it needs, and the fungus in turn provides the food needed by the ants. The fungus needs to be efficient in the event that in some years the amount of nourishing leaves may not be as abundant as in other years. In turn, the ants need to be efficient foragers of leaves for the same reason.

Parasitism

Parasitism is a relationship between two organisms in which one (the parasite) benefits from or lives off of the other, like fleas on a dog.

Ophiocordyceps unilateralis is an insect-infesting fungus, discovered by the British naturalist Alfred Russel Wallace in 1859 and currently found predominantly in tropical forest ecosystems. This micro catalyst "zombie fungus" infects mezzo catalytic ants with its spores, with the ultimate effect being alteration of the behavioral patterns of the infested ant. The fungus must attach securely to the ant's exoskeleton and penetrate it—avoiding or suppressing host defenses—then, control the behavior of the host before killing it; and finally, it must protect the carcass from microbial and scavenger attacks.

Infected ants leave their canopy nests and foraging trails seeking the forest floor where the temperature and humidity are suitable for fungal growth. The ant then uses its mandibles to affix itself to a major vein on the underside of a leaf until its eventual death in 4–10 days. Meanwhile, fungal fruiting bodies grow out of the ant's exoskeleton and release the fungus' spores.

O. unilateralis has been known to destroy entire ant colonies. In response, ants have evolved the ability to sense that a member of the colony is infested; healthy ants will carry the dying specimen far away from the colony in order to avoid fungal spore exposure. This fungus in turn is also susceptible to another fungal infection, which can limit the first fungus' impact on ant populations.

O. unilateralis found that ants provide it with the energy it needs to survive. The convenience of the abundant number of ants in a colony provided an efficient way to get to this source of energy.

Communities

Countries

When we look at a nation, we see resemblances to a living organism like the human body. It has a system of arteries (roads) that carry food, energy, and other goods to each cell (town), and a system of veins that carry away waste products away from each cell. It has an immune system that fights off threats to the "organism" (a defense force). It has a nervous system (government) that guides its actions and senses (news media) that work as its eyes and ears. It has a sort of DNA (a constitution and its laws) that help all other systems work. This "DNA" is packaged and dispersed to other parts of the globe in hopes that it takes root and grows (reproductive system).

When we see a town, we see a smaller version of a country where there is a constitution (charter or mission statement), a defense force (police), a local government, and ways to deliver food and water and to dispose of waste. It has a local newspaper to circulate information. Similarly, a company is a smaller version of a town, with goals and ways of bringing in resources to its individuals, removal of waste, and so on.

It is important to look at a country as a macro catalytic organism because this approach helps explain its evolution and behavior. Countries may seem to have personalities, like a next-door neighbor or a pet cat. These personalities develop from an amalgam of belief systems that have survived and prospered in ever-changing environments.

It can be added here that present-day countries exhibit varying degrees of energy efficiency. Affluent ones seem to be less efficient than others. This state of affairs is due to the abundance of fossil fuels rendering energy efficiency less than a critical factor.

Cycles

For the most part, Earth's matter moves in a circular fashion. For example, water in the ocean evaporates due to the heat of the sun and forms clouds that may travel to exposed lands. There, these clouds burst into rain creating streams of water that aggregate to create creeks and rivers that carry the water once again back to the ocean. This is an example of the hydrologic cycle of water on Earth.

Such cycles are the rule and not an exception for the movement of matter on Earth and are manifested in each species with a circulatory system. Groups of species such as those found in a forest circulate matter in the same way. One can say that energy drives the circulatory system of forests on Earth, and that without this circulatory system, forests would collapse because each of its links, be it another species or inanimate matter, is a tiny step in recycling of life-giving nutrients, making them available for regeneration of life and therefore maintaining the life of the forest.

When we look out a window or stroll through a park, we may see trees, birds, grass, and bees to name just a few of the life forms present. If we focus on a flower, many of us look at it as an object independent of the other objects around it, but in reality, it is intimately connected with all other objects in the picture we see.

Imagine that the flower we are looking at is clover. Clover needs soil to grow in, and as a mature plant, it provides nourishment for the bison roaming the Midwest plains. The puma on the other hand seeks the bison for food and when it manages to bring one down, it drags what it can of the carcass to the shade of a tree to eat. Flies gather around the carcass to feed and lay their eggs avoiding the birds that feed on them.

The above pictures and explanation are simplified, but they show how different species are interlinked. A more realistic representation of the link between species is circular as shown in the following graph.

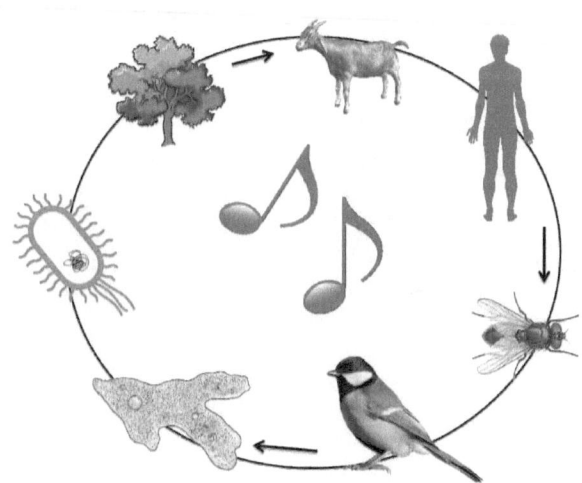

If we follow the biological cycle above clockwise starting with the green bush at the top left, we can tell a story of how nutrients flow from one live object to another. First, the bush extracts water and minerals from the soil and with the help of the sun and atmosphere produces tender leaves which are consumed by the goat to build its body and maintain its growth and health and to give birth to its offspring. The goat in turn is consumed by the human in the picture. The leftovers from the goat are used by the fly to lay its eggs. Some birds eat flies, and all of the preceding species—the goat, human, bird, and fly—as a result of digestion, leave behind fecal matter that is consumed by various microbes such as amoebas and bacteria, which return the nutrients (extracted from the soil by the green bush) to the soil once again.

Like the laws of physics that govern the universe in its grand symphony of reality, the musical notation in the preceding diagram of the living cycle signifies the rhythm with which this process is carried out on Earth.

On any territory, there are multitudes of living cycles but with different species that participate in returning the nutrients absorbed and transformed by the plants back to the soil and to the atmosphere. In addition, these living cycles can have relations with one another. For instance, a living cycle that starts with a bush can have more than one grazer consuming it such as goats, sheep, and deer. These living cycles can take the following shape:

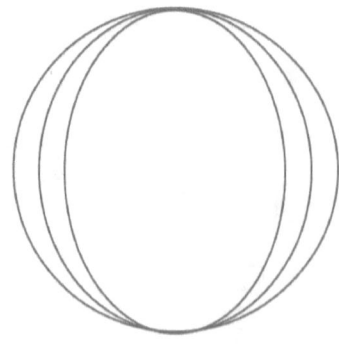

If we look at a square acre of land and try to describe all the living cycles within it, we may come up with a million combinations, some of which may look like the living cycles in the next diagram:

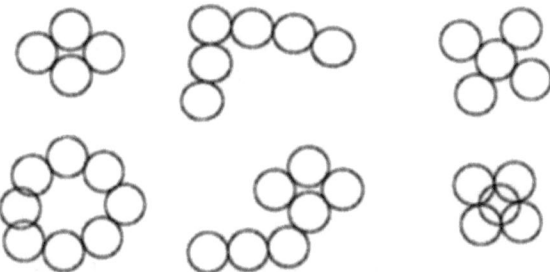

Living cycles are not limited to sources of food. They can be related to nest building and so on. The biological cycle graphic below depicts a bird building a nest in a tree. When the eggs hatch, the nest may fall to the ground where insects like termites break it down passing on the resulting matter to amoebas and bacteria resulting in the return of the nutrients back to the soil.

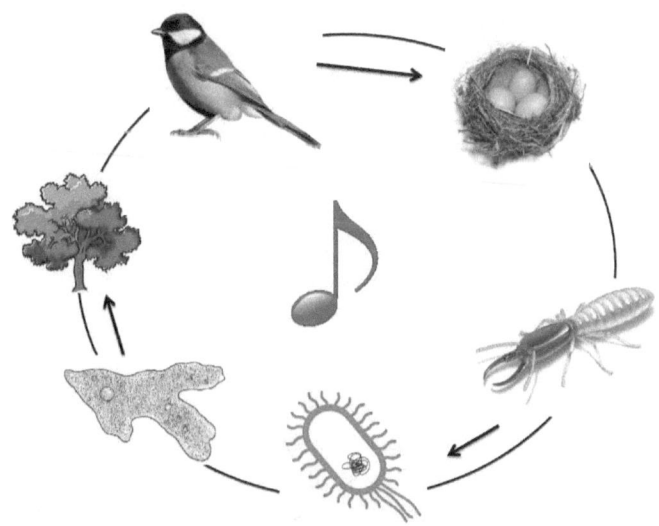

This is probably a good point to introduce the concept of an ecosystem. A forest is an example of one. An ecosystem is a mega catalyst that describes the relations or circular links between air, soil, water, plants, and animals of a forest. A forest is not the only ecosystem. A desert such as the Sahara is also considered an ecosystem as well as the biota of a continent.

Cycle Stability

At this point, with respect to the living cycle, one might argue that this cycle is unstable. For example, a gazelle herd can consume all the vegetation in their territory then starve to death the next week or year. At the same time the living cycle of the gazelle also dies because moving clockwise in the diagram below, all subsequent living things will also starve.

But we know that gazelle graze off of the vegetation in their territories without consuming it all. We may ask the question, "How do they temper their appetites so that they do not exhaust the vegetation?"

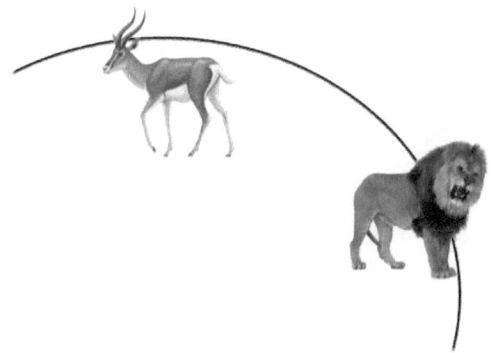

When we look at the relationship between the lion and gazelle, the African lioness eats from the herd of gazelle to satisfy her hunger and the hunger of her pride. She does this in a way that always leaves enough gazelle for her to hunt the next day or the next year. If she finished off the whole herd of gazelle, then the lioness' cubs will have nothing to eat the next day or year and die off, the end result being a broken living cycle.

One key factor that gives stability to this living cycle is laziness. In the wild, it has been observed that lions sleep a lot under the shade of trees and go out hunting for gazelle or other prey only when their bellies rumble; to conserve energy lions go after slow individuals in the gazelle herd. If laziness was not present in the pride of lions, one can imagine the lionesses hunting all day and bringing down as many gazelle as they could, which would in the end break the living cycle.

Another key factor is territoriality. Lions are known to have territories that keep other hungry lion prides out and therefore preserve the herd of gazelle from overhunting. At the same time, the gazelle are involved in the preservation process too. They have evolved the ability to flee when the hunt is on. . In addition the gazelle have a mechanism that keeps them from over grazing.

Thus, with the lion helping to control the number of individuals in the herd of gazelle, the gazelle in turn tend not to overconsume its food source. All these physical and behavioral characteristics are genetically controlled and evolve to help stabilize the living cycle.

Another factor in living-cycle stability is the availability of substitute species. For example, if a fly—that uses the lion pride's leftovers of food as a habitat for their larvae—goes extinct, then another species will take over as a substitute in processing these leftovers so that the living cycle is not broken.

Then there are the generalist species of flies that use leftovers from several other predator species. For instance, in an ecosystem, there may be a fly that can lay eggs on leftover carcasses of prey of the lion, cheetah, hyena, or wild dogs.

In summary, living cycles are stable over time because the genetics of their participants have adapted to self-preservation in the course of evolution.

Cycle Instability: The Algae Bloom

The term "bloom" refers here to the explosion of algae density in a pond. To understand what an algae bloom does, we first must understand how a pond of water operates. A pond naturally occurring in a forest is composed of a pond bottom consisting of soil or clay that is a depression in the ground filled with water. On top of the water is air. Algae grow in water and receive sunlight from above and nutrients from the water and the soil beneath. As the algae grow, mature, and die, they sink to the bottom of the pond. There, bacteria decompose the algal matter and release key elements that then circulate in the pond water nourishing subsequent generations of algae.

Algal blooms in ponds are common. They occur when the conditions are right for algae to undergo a sudden or prolonged burst of growth. This burst is usually due to an increase in one or more chemical resources in the water essential for algal growth such as compounds containing phosphate.

Most blooms are benign toward the water system. The water may turn bright green for a few days or a week or so. The algae that bloomed die and sink to the bottom of the pond when the chemical or chemicals that initiated the bloom are exhausted.

Some water systems experience long blooms. These may last from a month to two or longer so long as the chemical resources responsible for the bloom continue to come into the pond water system, such as an effluent from sewer systems. This process can be devastating to the water system's biota and cycles because the increase in the number of alga corpses being consumed by bacteria (which are oxygen consumers) will cause the lake to become oxygen deficient for many plants and animal species, which in turn will die off. This type of bloom chokes life out of the lake, which will eventually die, culminating in anaerobic bacteria's jumping into action, producing foul smells like rotten eggs associated with stagnant water systems, after most if not all oxygen is consumed.

You would think that even if the chemical sources that triggered the bloom in the first place continue to flow into the pond, the bloom will continue indefinitely, but on the contrary, the blooming species will eventually die off and sink to the bottom of the water system if other chemical resources necessary for life are depleted as a result of the bloom. You see, algae depend on many chemical resources present in the water system and when one or more of these are depleted, the bloom cannot continue, and the species in question dies off sinking to the bottom, where the bacteria go to work.

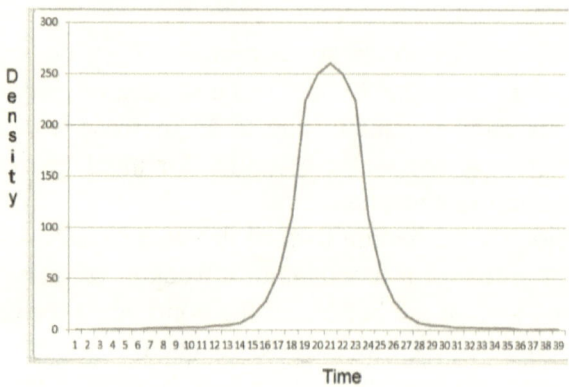

The density of algae in a pond undergoing a long bloom

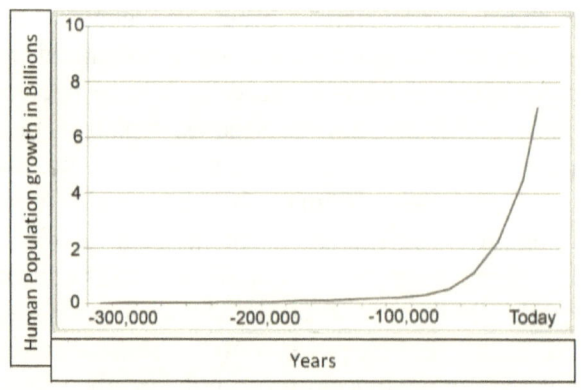

Is Earth Undergoing a Human Bloom?

To bring back a pond that has experienced a bloom, we need to take out the algae, plant matter, and animals that died during the bloom and the associated decomposition-related bacteria. In addition, all sources of chemicals responsible for the bloom must be stopped from entering the pond. Only then will enough oxygen be reabsorbed by the pond from the air above, allowing fish, plants, and algae to grow once again and continue the life cycle of the pond.

You may ask, "How does the explanation of pond blooms relate to the formation and stability of cycles?" Mankind is believed to be undergoing a "bloom" that carries with it a mass extinction of 50% or more of all life on Earth. Underlying this argument is the idea that we as humans and our habitat are no different from the algae in the pond. It's all a matter of chemical reactions that follow physical laws over which we apparently have no control.

In these modern times, we face a whole new set of challenges that differ from the challenges faced by our Greek ancestors such as Socrates and Plato. In those early times, the challenges were those of aggregating people together into cities that are under the control of fair and acceptable governing bodies and laws. Today, we face a whole new set of problems related to the biological "success" of our species.

The Niche

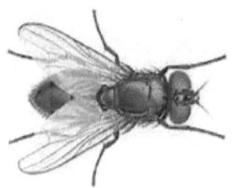 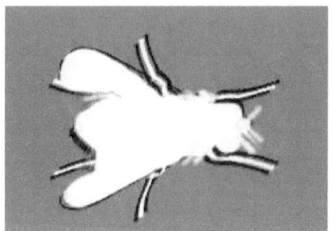

The niche of the fly is everything necessary for the survival of the fly that is not the fly itself. For example, the image to the left above is of the fly and not its niche. The gray image to the right of the fly is everything else on which the fly depends minus the fly itself and is called its niche. The gray area is the air the fly depends on for flight as well as breathing, the food that it requires, the material in which it lays its eggs, the temperature of the surroundings where it goes to sleep at night, the perch on which it stands, the fly that fertilizes its eggs, and so on.

The niche is also the environment in which the fly lives most efficiently. You could say that the niche is a perfectly tailored suit for the fly. For example, a fly that is used to live in the tropics will die in the polar areas because of the temperature. A tropical fly that is transported to a temperate region will not survive either because it will not find the right food or shelter. A tropical fly that is moved from the tropics of Africa to the tropics of Central America will probably have a harder time being as efficient as the local African flies.

Therefore, the suitability of a niche is important to the species and in turn to the living cycle in which it serves as a conduit of energy and other resources. For instance, in the diagram above, if the environment changes rendering the bird's niche harder to live in, the birds will not be as efficient at controlling the population of flies, and therefore the flies will multiply faster, and in the end, be a nuisance for the gazelle and lions, who will in turn waste more energy and time trying to chase them away or avoid the fly hoards, to say nothing of the fly's ability to spread a disease disrupting the living cycle.

To recap, the niche comprises all the resources required to fulfill an individual's livelihood and role in the balance of the living cycle. The niche of the gazelle is a part of the niche of the lion, and the niche of the lion is a part of the niche of the fly, and so on.

Hassan Rasheed

Succession of Mega Catalytic Ecosystems

An ecosystem is a collection of species that are interdependent on each other and on the nonliving surroundings in which they evolved such as mountains, soil, water, and energy. We can also call an ecosystem a unique mega catalytic organism because its cooperating individuals include large numbers or plant and animal species. Ecosystems go through a process of environmental succession. This succession is driven by the ever-changing conditions on Earth such as climate and terrain. For example, with respect to climate, there were ice ages at different periods in the ancient past. Some territories were once deserts that are now lush forests and vice versa. As for terrain, whole continents are constantly moving around the globe, volcanoes belch up magma from deep underground, rain pulverizes rocks converting them to soil, and so do certain plants.

These changes in turn alter the relationships between species as well as the types of species present in an ecosystem. For example, if we start with a rocky terrain we often find that the living cycles present are simple ones and involve miniature species. If you look at a rock in a temperate zone of the globe, you may find more often than not lichen growing on it. Like any species, lichen grows and dies. Its remains are then consumed by various insects and microbes, and the nutrients it was able to absorb from small cracks in the rock are released into the environment once again.

As the lichen grows and creates more cracks in the rock's surface, the rock starts to flake away, and the flakes accumulate creating soil that other plants, such as thistles and legumes, can use to grow. This process creates a new type of living cycle that involves herbivores like the bison, carnivores like the puma, and parasites like flies. Not only are the species involved larger in size than those in the lichen living cycle, the number of species involved also grows. For example, the pictures below show this transition from left to right of a simple living cycle to a more complex one. This is called ecosystem succession.

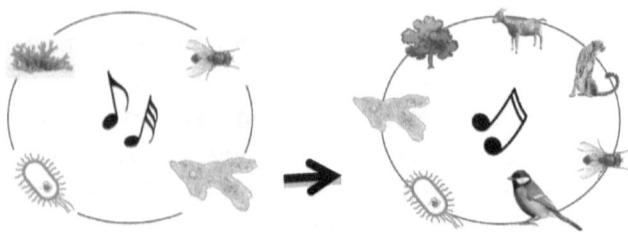

You should realize that the start of the living cycles above are the plants that take up minerals and water from the soil, carbon dioxide from air, and energy from the sun, thus producing living matter and in turn feeding the other species in each living cycle. These plants also create whole three-dimensional environments above ground as the succession process continues into forests and jungles containing thousands of living cycles and millions of species.

The new species that fill these new ecosystems are generated by evolution.

Maturation

Evolution operates on a trial-and-error basis. Those born with an error or sub performing trait perish, whereas those born with a successful trait survive. So, it is with the relationships of species that belong to a cycle. If a species does not fit properly in its cycle, then it will die and be replaced by one that is a better fit. This process continues until the cycle is most stable.

For example, when a mezzo catalytic African lion consumes a gazelle, it squeezes most of the energy out of the gazelle's flesh to maintain itself, produce young, and then it leaves the remainder of its digestion on the ground. This remainder is then consumed by insects, insect larva, and a host of microbes that squeeze even more energy out of what that lion left behind. This process releases critical minerals into the ground, making them available to the grass and legume seeds and plants that in turn use them to convert the sun's energy into lush vegetation.

This vegetation is then consumed by the gazelle and the cycle continues, the end result being zero impact, where a few critical minerals are transported up the stalks of grasses and legumes with the help of the sun, then down to the ground once again through the gazelle, lion, insects, and microbes.

Lions spend much of their time resting and are inactive for ~20 hours a day. Although lions can be active at any time, their activity generally peaks after dusk with a period of socializing and grooming. Intermittent bursts of activity follow through the night hours until dawn, when hunting most often takes place. They spend an average of two hours a day walking their territories and 50 minutes eating.

The size of adult lions varies across their range. In Zimbabwe, males average approximately 418 lbs. (280 lbs. for females), while in East Africa, males average 386 lbs. and females 263 lbs.

Lionesses do most of the hunting for their pride. They are more effective hunters because they are smaller, swifter, and more agile than the males and unencumbered by the heavy and conspicuous mane, which causes overheating during exertion. They act as a coordinated group with members performing the same role consistently in order to stalk and bring down the prey successfully. Smaller prey is eaten at the site of the hunt, thus being shared among the hunters; when the kill is larger, it often is dragged to the pride's area. There is more sharing of larger kills although pride members often behave aggressively toward one another as each tries to consume as much food as possible. Near the conclusion of the hunt, males have a tendency to dominate the kill once the lionesses have succeeded. They are more likely to share this food with the cubs than with the lionesses, but males rarely share food they have killed by themselves.

The African lioness eats from the herd of gazelle to satisfy her hunger and the hunger of her pride. She does this in a way that always leaves gazelle for her to hunt the next day or the next year. You might ask, "How does she do that?" The genetics of the lion has matured over time and ensures conservation; if a lioness finished off the whole herd of gazelle, then her cubs would have nothing to eat the next day or year and die off, the pride of lions would go extinct.

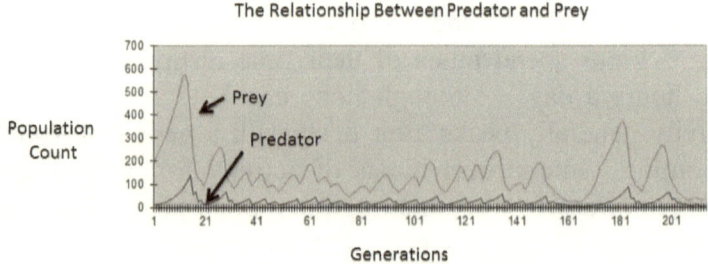

The above graph shows a stable relationship between a predator and its prey over 210 generations. The fluctuation in the population count accounts for perturbations in the environment such as years when the foliage that the gazelle feed on is scarce or abundant.

Summary
There should be no doubt that any one catalytic organism is connected in some way or another to every other creature in an ecosystem through the cascade of sun's energy from plants to the rest of the biota. It should also be clear that no one creature or cycle is more important than another in an ecosystem community. In addition, the sun's energy is a limited resource, and life forms, strive to use it in the most efficient way possible.

Humans

Human biological evolution is the process of change by which people originated from apelike ancestors. One of the earliest defining human defining traits is the ability to walk on two legs, which evolved over 4 million years ago. Other important human characteristics such as a large and complex brain and the capacity for language developed more recently. Many advanced traits including complex symbolic expression, art, and elaborate cultural diversity emerged mainly during the past 100,000 years.

The evolution of bipedalism, a larger brain, and the capacity for language all contributed to the efficiency with which these early human catalysts acquired the scarce energy needed to survive.

Humans are primates. Physical and genetic similarities show that the modern human species, *Homo sapiens*, has a very close relation to another group of primate species, the apes. Humans and the great apes of Africa (chimpanzees, baboons, and gorillas) share a common ancestor that lived between 8 and 6 million years ago. Humans first evolved in Africa, and much of human evolution took place on that continent. The fossils of early humans who lived between 6 and 2 million years ago come entirely from Africa.

Currently, approximately 20 species of early humans are believed to have existed. Scientists do not know how some of these human species are related. Many early human species left no living descendants. Scientists also debate how to identify and classify particular species of early humans, and what factors influenced the evolution and extinction of each.

More important is the fact that humans like all the life forms on Earth owe their lineage to the very first living entity that must have appeared on Earth 3 or 4 billion years ago. This lineage carries with it the ability to use cooperation, symbiosis, parasitism, and a host of other traits to get established in a niche belonging to a cycle or cycles in an ecosystem.

This human niche was one of many participants in the efficient use of the sun's limited energy that reached the Earth where its capture started with plants and similar organisms. Once captured, it cascaded down through each ecological cycle from a prey to predator, from a host to parasite, or was shared in a symbiotic relationship to name just a few.

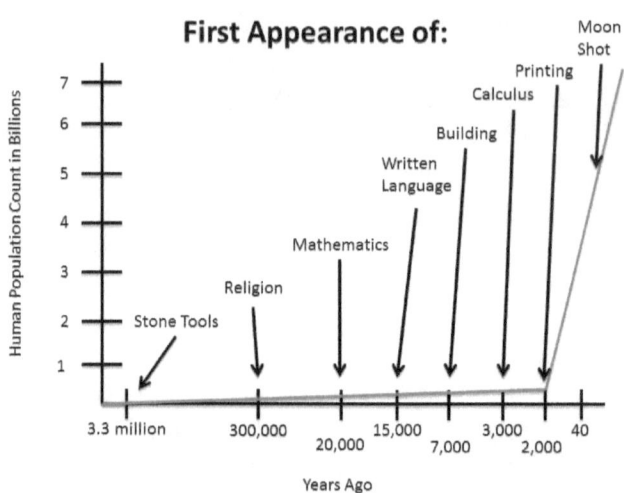

Approximately 200–300 years ago, humans broke away from their natural niche and the associated cycles with the discovery of the availability of (what seemed limitless) energy from fossil fuels. The result was the industrial revolution. Humans, like all other creatures, behave as if they are the most important being on Earth. All the living must assume that they are at the center of the universe because any other way they would have no motivation to continue.

With this new power source and motivation, humans set out to control their own destiny and proceeded to tame the world. Instead of a cyclical existence they felt comfortable with a straight line. So they built square homes, towns, and cities. They denuded the jungle and built farms with machinery that could only go in one direction and so farmland was divided into square or rectangular fields.

A natural jungle

Human landscape

Instead of being a part of the circular food chain, they chose to be at the top of a food pyramid. If a creature was thought to be useful to human existence it was domesticated or farmed. If it was not thought of as being useful to humans, it was ignored or eliminated.

Instead of being more efficient in the use of energy, humans became less so. They built their own monetary systems based not on the value of energy for the sustenance of life but on the commodity value of material things (see Appendix 1). Energy was no longer a precious resource but a way to heat up monetary economies to support material-based growing populations. Instead of being a part of the biota, they chose to be apart from it.

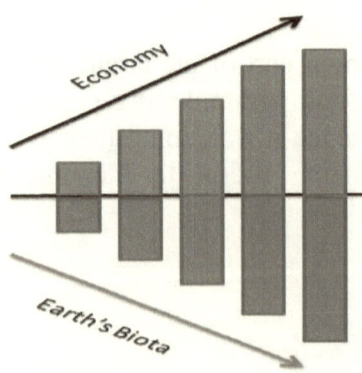

The task in front of us now is to find out how to abate this human bloom, but first we need to discuss its dark side effects and the roadblocks to really reverse this disaster.

Darkness

A person cannot help but acknowledge that evolution has done "well" with the human species. Humankind has successfully competed with its neighboring species leaving more offspring in subsequent generations and grew exponentially. There have been, however, indications that not all is well with this human bloom (please refer to the reading list on the prospects of the future at the end of this book). We got lost from our original physical and behavioral selves and that will be discussed in later chapters, but presently we need to talk about our effects upon the environment.

The following are but a few important examples that have popped up in the news in the past few years regarding the negative effects of the explosion of humankind's population upon the planet Earth's infrastructure; these effects destroyed multitudes of Earth's energy-efficient catalytic cycles:

Coal and Crude Oil

The use of coal and crude oil greased the skids powering the industrial revolution and every conceivable aspect of modern life from fuel for mechanical engines to plastics that seem to permeate and envelope everything we buy. This use of coal and crude oil also hailed in the haphazard research in chemistry where new and nondegradable and nonbiological chemicals now poison our existence from killing precious pollinating bees to seeping into our water and food systems.

Crude oil is so pervasive because it is the lifeblood of our economies. There are three issues concerning crude oil and its products that affect the environment. The first is air pollution mainly from automobile exhaust fumes. These pollutants come from total and partial combustion of fuels. They include carbon monoxide and carbon exhaust particulates.

Animals exposed to high enough concentrations of certain air pollutants may experience
- Irritation of the eyes, nose, and throat
- Wheezing, coughing, chest tightness, and breathing difficulties

- Worsening of existing lung and heart problems, such as asthma

- Increased risk of a heart attack

In addition, long-term exposure to air pollution can cause cancer and damage to the immune, neurological, reproductive, and respiratory systems. In extreme cases, it can even cause death.

The second issue is the spilling of crude oil and its byproducts because of human activity. Approximately 11% of these spills occur during the drilling, refining, and transportation of these products; 89% are caused by the end user in the dumping of petroleum-based products down the drain in our homes and businesses, leaky engines and mishandling of fuels in activities such as mowing the lawn or faulty storage tanks that leak into the ground.

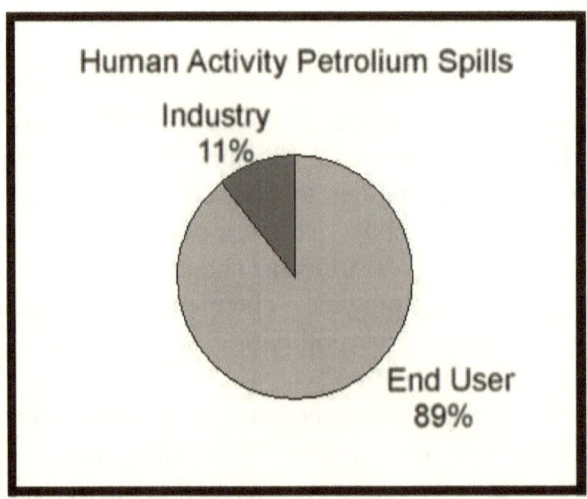

A graph showing percentages of spills due to petroleum industry and end user activities

Third, crude oil, refined petroleum products, and polycyclic aromatic hydrocarbons are ubiquitous in various environmental zones. They can bio-accumulate in food cycles where they disrupt biochemical or physiological activities of many organisms, thus causing carcinogenesis of some organs, mutagenesis in the genetic material, impairment of reproductive capacity and/or hemorrhages in exposed populations.

Tourism

Idyllic tourist destinations have been so spoiled by tourists that many authorities are being forced to close them down to prevent further damage.

For example, Koh Tachai in Thailand's Ranong Province was rated as Thailand's most beautiful island. Nevertheless, according to the Bangkok Post, the tourist-ravaged beaches of Koh Tachai will have to close indefinitely.

Tunya Netithammakul, director general of the department of national parks, wildlife, and plant conservation, said that tourism had resulted in "overcrowding and the degradation of natural resources and the environment."

The closure of the island was designed to give the land and marine environments a chance to regenerate "before the damage is beyond repair."

Mining

The environmental impact of mining includes erosion, formation of sinkholes, loss of biodiversity, and contamination of soil, groundwater, and surface water by chemicals from mining processes. In some cases, additional forest logging is done in the vicinity of mines to increase the room available for storage of the created debris and discarded soils. Besides causing environmental damage, the contamination resulting from leakage of chemicals affects the health of the local population. Mining methods have significant environmental and public health issues.

Erosion of exposed hillsides, mine dumps, tailings dams, and the resultant siltation of drainages, creeks, and rivers can significantly impact the surrounding areas, a prime example being the giant Ok Tedi Mine in Papua New Guinea. In areas of wilderness, mining may cause destruction, disturbances of ecosystems and habitats, and in areas of farming, mining may disturb or destroy productive grazing and croplands. In urbanized environments, mining may produce noise pollution, dust pollution, and visual pollution.

Mining refineries produce a large proportion of the harmful waste products, which later have to be disposed of in specific ways in order to avoid harming the environment. In addition, conventional refineries create their own toxic waste that must also be accounted for.

Governments around the globe that have some kind of "environmental" concerns and attempt to regulate mining do it to protect human safety, and in some instances, the safety of human properties not the natural fauna and flora. No concept of the cyclical nature of life on Earth is understood.

Agriculture

From the birth of agriculture, food production for humans has sequestered the natural resources of other species for human use to the point where humanity is faced with the task of becoming zookeepers of what remains of Earth's observable wildlife. The massive human population explosion is not helping but hindering the work of those who see its danger to the planet because when it comes to the last kernel of maize, the odds are that it will be given to a human and not a bird or squirrel.

Agriculture, as with any perturbation to the natural environment, has also interfered with evolution and natural selection, where naturally occurring plants and animals are replaced with domesticated ones. For instance, in the process of farming, our ancestors had a choice of picking from two varieties of wheat they collected on the hillsides of the Near East. The dominant type shattered upon the slightest touch, scattering the seeds so the species could spread and survive. The other did not scatter its seeds so easily, and therefore it was easier to handle. As a result, a higher proportion of this non-scattering variety was collected and planted than occurred in nature. With each succeeding year, a higher proportion of the non-scattering wheat was harvested and planted. Natural selection had been replaced by those who think they know best.

With agriculture, there were times when production exceeded demand, and as a result, politics was born; a decision as to who gets the surplus and what they do with it came into play, and a new world of wheeling and dealing came into being.

In summary, humans once danced to the beat of the whole Earth in unison with all life forms, as hunter-gatherer's living cycles contributed to the web of life. It is true they took but they also gave back and never took for granted what they received, and instead believed that it was a precious gift to be thankful for. The Earth was a part of them, and they were a part of it physically and spiritually.

Unfortunately, the thoughts of some humans wandered beyond the planting of seeds. One of the first wayward thoughts included domestication of animals. You may argue that it does not matter if you eat from a wild animal or a domesticated one; there is no difference pound for pound.

The breakthrough of the unconstrained Earth dancers did not stop there. The humans needed to feed their captives, and so they had to impound the grounds that others danced on. They denied those others the right to feed freely. Plants, birds, and insects were eliminated or chased away and so were the many animals that called these lands their temples.

Pesticides

Through most of history, insects have been loathed. From locusts to termites and the boll weevil they have gained a reputation as being detrimental to the interests of humankind. It was not until recent history that some insects were found to be useful and helpful to humankind.

Over 98% of sprayed insecticides and 95% of herbicides cause harm to life forms other than intended target species because they are sprayed or spread across entire agricultural fields. Runoff can carry pesticides into aquatic environments, whereas wind can carry them to other fields, grazing areas, human settlements, and undeveloped areas, potentially affecting other species. Other problems emerge from poor production, transport, and storage practices. Over time, repeated application increases pest resistance, while its effects on other species can facilitate the pest's resurgence.

Each pesticide or pesticide class comes with a specific set of environmental concerns. Such undesirable effects have led many pesticides to be banned, while regulations have limited and/or reduced the use of others. Over time, pesticides have generally become less persistent and more species-specific, reducing their environmental impact. In addition, the amounts of pesticides applied per acre have declined, in some cases by 99%. Nonetheless, the global spread of pesticide use, including the use of older or obsolete pesticides that have been banned in some jurisdictions, has increased overall.

The bee is just one example of the harm done to off-target species. It was not until its effects on pollination were discovered that value of bees' services to humankind was recognized along with that of several other insects. Today, bees and other useful insects are in danger from insecticides that are used to protect domesticated plants from other insects and fungi.

Fertilizers

Similarly to insecticides, fertilizers can run off into waterways thereby poisoning fish and other aquatic species for which fertilizers were not intended.

With the advent of the so-called Green Revolution in the second half of the 20th century—when farmers began to use technological advances to boost yields—synthetic fertilizers, pesticides, and herbicides became commonplace around the world not only on farms but also in backyard gardens and on front lawns.

These chemicals, many of which were developed in the lab and are petroleum-based, have allowed farmers and gardeners of every stripe to exercise greater control over the plants they want to grow by enriching the immediate environment and warding off "pests." But such benefits haven't come without environmental costs—namely, the wholesale pollution of most of our streams, rivers, ponds, lakes, and even coastal areas—because these synthetic chemicals run off into nearby waterways.

When the excess nutrients from all the fertilizers we use enter our waterways, they cause algae blooms sometimes big enough to make waterways impassable.

A related issue is the poisoning of aquatic life. According to the U.S. Centers for Disease Control and Prevention (CDC), Americans alone churn through 75 million pounds of pesticides each year to keep the "bugs" off their peapods and petunias. When those chemicals get into waterways, fish ingest them and become diseased. Humans who eat diseased fish can themselves become ill, completing the circle wrought by pollution.

A 2007 study of pollution in rivers around Portland, Oregon, revealed that wild salmon there are swimming around with dozens of synthetic chemicals in their system. Another recent study from Indiana showed that a variety of corn genetically engineered to produce the insecticide Bt is having toxic effects on off-target aquatic insects, including caddis flies, a major food source for fish and frogs.

Economics

Any modern economy is fueled by cheap energy scraped from or pumped out of the ground, and when the goal of this economy is to provide endless goods and services to its limitlessly self-gratifying consumers, the finite Earth must pay the price with gored landscapes and poisoned waterways and agricultural fields. This open-ended perception that the Earth is limitless in providing resources is a problem we need to face today and not tomorrow.

Extinctions

The fastest mass extinction in Earth's history is happening right now.

There is no doubt in the minds of biologists that the Earth is losing a race with time to save its precious cargo. In 1993, Harvard Biologist E.O. Wilson estimated that Earth was losing ~30,000 species per year or 3 species per hour. This means that approximately 700,000 species have gone extinct since this estimate was made. This bio-die-off crisis is snowballing, and many experts predict the sixth mass extinction on the horizon.

There is no doubt that humans are the direct cause of this habitat stress and species destruction in the modern world through such activities as transformation of the landscape (through agriculture, mining, oil exploration and fracking, construction and forestry practices), overexploitation of species (by overhunting game species), pollution, human overpopulation, and introduction of invasive species into existing habitats.

Food Chain Disruption: The most immediate impact would likely be on the food chain. For example, as sea surface temperatures continue to rise, many species of plankton are beginning to decline. If the plankton, such as diatoms and krill, were to go entirely extinct, this event would have an impact on larger creatures, such as fish and whales, who consume plankton as a major food source. In turn, if those larger marine animals have less to eat, and as their own population declines, these changes may have a domino effect throughout the food chain, ultimately reducing human food sources.

Medicine: Many medications are derived from plants, which rely on insects for pollination. As the insect population decreases, such plants may struggle to reproduce. The fewer plants there are, the harder it would be for the remaining insects to find a food source, creating a cause-and-effect loop that can hurt all the species involved. Some medications are created from the skin of amphibians, which are also under threat of extinction.

Higher Disease Risk: Some animals are resistant to certain types of diseases, e.g., the opossum's ability to ward off Lyme disease. Such buffer species help contain outbreaks from spreading to other animals and humans. As buffer species lose habitat space to urbanization or as climate-induced changes in an ecosystem proceed, the ability of such species to create the buffer declines. As they become extinct, the creatures that move in to take their place are often less adept at containing the spread of diseases or are more likely to contract such diseases, putting humans at a higher risk of diseases.

The Habitats

With the domestication of plants and animals, more and more lands are needed to support this human activity, implying a loss of a habitat for animal and plant species of no interest to humans.

For instance, in the southern United States of America, beef cattle are raised on ranges making them compete with other indigenous species that naturally occur there. This competition is aided by humankind's ability to shoot or poison those species that are undesirable to the national economy.

Soil Erosion and Degradation

Natural processes moved rocks from their resting places down hills and gullies using wind, whereas water currents caused them to collide with other rocks. All this movement resulted in the formation of soils approximately 450 million years ago, when plants started to take root after evolving from simpler organisms. Since then, soils have moved under the influence of water or wind (known as erosion) at a comparable rate of formation and deposition.

A far more recent problem is the loss of soil at a much faster rate than its formation or deposition from other areas. This phenomenon results from humankind's actions such as allowing overgrazing by cattle and other domesticated species and improper cultivation practices by means of mechanized agricultural implements. These actions leave the land fallow and unsuitable for human or natural-species use. The eroded soils are transported away and deposited in undesirable locations.

As a result of human construction activities, soils may also be degraded, which includes salinization, nutrient loss, and compaction.

Approximately 40–50% of the soil erosion on Earth is caused by humans and their irresponsible behaviors. Deforestation, improper farming techniques, intensive agriculture, and construction of roads, bridges, and dams all lead to soil erosion.

Although it is true that intensive agriculture and improper farming methods lead to soil erosion, it is equally true that soil erosion affects further agriculture. When the soil is being eroded, it is not just the soil that is taken away but the minerals and nutrients that naturally occur in the soil are also lost. Soil erosion removes the topmost layer of the soil (topsoil), which no doubt, is the most fertile and most productive part of any soil. When the top layer is removed, rills and gullies are formed in the soil.

The harmful effects of the removal of topsoil do not end there. When the top layer is removed, the soil's ability to store water and other nutrients are reduced. This situation also exposes the subsoil. This subsoil has poor physical and chemical properties making it susceptible to further erosion, and crops that were newly planted are washed off along with the soil.

This soil erosion also affects waterways. It leads to deposition of silt in the path of flowing water. The eroded soil may contain fertilizers, pesticides, and other harmful chemicals, which will decrease the quality of water in these rivers and streams.

Soil erosion compromises the soil's capacity for sustaining the growth of plants and crops; thus, agricultural yields tend to be lower. Often, the subsoil contains clay, which limits the development of the roots of plants.

Wind erosion contributes to air pollution. Soil particles carried in the air lead to dust that can contain chemicals from agricultural practices. This soil dust often leads to respiratory problems and skin infections in humans.

Ocean Pollution

This pollution occurs when harmful or potentially harmful effects result from the entry into the ocean of chemicals, particles, industrial, agricultural and residential waste, noise, or the spread of invasive organisms. Eighty percent of marine pollution comes from land. Air pollution is also a contributing factor by carrying off pesticides or dirt into the ocean. Land and air pollution have proven to be harmful to marine life and its habitats.

The pollution often comes from nonpoint sources such as agricultural runoff, wind-blown debris, and dust. Nutrient pollution, a form of water pollution, means contamination by excessive input of nutrients. It is a primary cause of eutrophication of surface waters, in which excess nutrients, usually nitrogen or phosphorus, stimulate algal growth.

Many potentially toxic chemicals adhere to tiny particles, which are then taken up by plankton and benthos animals, most of which are either deposit feeders or filter feeders. This way, the toxins are concentrated upward within ocean food chains. Many particles combine chemically in a manner highly depletive of oxygen, causing estuaries to become anoxic.

When pesticides are incorporated into the marine ecosystem, they quickly become absorbed into marine food webs. Once in the food webs, these pesticides can cause mutations, as well as diseases, which can be harmful to humans and to the entire food web.

Toxic metals can also be introduced into marine food webs. These metals can cause a change in tissue composition and biochemistry and in animal behavior and reproduction, and thereby suppress growth in marine life. Also, many animal feeds have high fish meal or fish hydrolysate content. This way, marine toxins can be transferred to land animals, and appear later in meat and dairy products.

Deforestation

It has been estimated that approximately a half of Earth's mature tropical forests—between 2.9 million and 3 million square miles of the original 5.8 million to 6.2 million square miles that until 1947 had covered the planet—have now been destroyed. Some scientists have predicted that unless significant measures such as seeking out and protecting current old-growth forests are taken on a worldwide basis, then by 2030, there will only be 10% remaining, with another 10% in a degraded condition; 80% will have been lost and with them hundreds of thousands of irreplaceable plant and animal species (along with a loss of habitat for millions of plant, animal, and insect species).

Forests are cut down for many reasons, but most of them are related to money or to people's need to provide for their families. The biggest driver of deforestation is agriculture. Farmers cut forests to gain more room for planting crops or grazing livestock. Often, many small farmers each clear a few acres to feed their families by cutting down trees and burning them in a process known as "slash and burn" agriculture.

Logging operations, which provide the world's wood and paper products, also cut countless trees each year. Loggers, some of them acting illegally, also build roads to access more and more remote forests; these actions lead to further deforestation. Forests are also cut as a result of urban sprawl.

Deforestation has many negative effects on the environment. The most dramatic impact is a loss of habitat for millions of species. Seventy percent of Earth's land animals and plants live in forests, and many cannot survive the deforestation that destroys their homes.

Deforestation also drives climate change. Forest soils are moist, but without protection from sun-blocking tree cover, they quickly dry out. Trees also help perpetuate the water cycle by returning water vapor back into the atmosphere. Without trees to fill these roles, many former forest lands can quickly become barren deserts.

Removal of trees deprives the forest of portions of its canopy, which blocks the sun's rays during the day and holds in heat at night. This disruption leads to more extreme temperature swings that can be harmful to plants and animals.

Trees also play a critical role in absorption of greenhouse gases that fuel global warming. Fewer forests mean larger amounts of greenhouse gases entering the atmosphere—and increased speed and severity of global warming.

Depletion of Global Fisheries

A report entitled *"Status and Solutions for the World's Unassessed Fisheries"* by the Sustainable Fisheries Group (SFG) has confirmed the suspicions held by many researchers that nearly 80% of the world's fisheries are in steep decline. The reasons for this decline are overfishing caused by many factors that include uninformed political pressure that tends to dominate the decision-making process, the problem of the unmanaged commons where there is no policy for management, and the use of gill nets that kills much more than the target fish. Over time, this situation can lead fisheries to collapse.

Overfishing is a global problem with many serious social, economic, and environmental implications.

Environmental Effects

There is also growing evidence that the increased volume of fishing activity worldwide is having a serious effect on the health of the oceans as a whole. When commercially valuable species are overexploited, other species and habitat that share the same ecosystem are affected.

For instance, recent studies suggest that overfishing of large shark species has had a ripple effect in the shark's food chain, increasing the number of species, such as rays, that are usual prey for large sharks, resulting in declining stocks of smaller fish and shellfish favored by these species.

In addition to harvesting large amounts of fish and seafood for sale, large-scale fishing operations catch and often unintentionally kill off-target marine life, including juvenile fish, corals, and other bottom-feeding organisms, sharks, whales, sea turtles, and birds. Killing these unintended species can have significant effects on marine ecosystems.

On the basis of new information about the dynamics of marine ecosystems, more and more countries and regional fisheries management organizations (RFMOs) are adopting an ecosystem-based approach to the management of fish stocks.

Invasive Species

These are species that are transferred from one ecosystem to another and compete with native ones by taking over their habitats. An increase in global trade has increased the number of plant and animal species that are carried from one part of the globe to another legally, illegally, or accidentally. Some of these species become invasive and wreak havoc on the destination ecosystem's ecological balance.

A key factor that makes many species invasive is a lack of predators in the new environment. This factor is complex and results from thousands of years of evolution in a different place. Predators and prey often coevolve in a phenomenon called the coevolutionary arms race. What this means is that as prey develop better defenses, predators in turn evolve better ways of exploiting prey. The classic example of this interplay is the cheetah and antelope. Faster antelope survive better because they can better escape cheetahs. The fastest cheetahs then survive better because they can better catch the faster antelope. Neither species ultimately gains an advantage because they continually evolve in response to one another.

Nevertheless, when a plant or animal enters a new environment, they will likely encounter predators who have not been evolving with them, which makes these predators unable to successfully exploit the prey. Defense mechanisms like venom, size, or speed that have been matched by adaptation in predators are suddenly without match in the new environment.

This advantage can allow the species to proliferate rapidly because it no longer faces any predators. Many insect or fungal species that are invasive in the United States come from regions where native trees have evolved resistance to their effects. When these species enter the US, they find trees that have no resistance and can decimate forests quickly.

Invasive species may also be able to exploit a resource that native species cannot use, which allows them to take hold in the new environment. Introduced into the Western United States, the barbed goatgrass thrives in serpentine soils, whereas native plants do not normally grow there. This exploit has given the barbed goatgrass a solid stronghold in the area. Combined with the fact that grazing animals do not like its taste, this grass has spread rapidly throughout California.

Some species also alter the environment in a manner that makes the environment more favorable for them but less suitable for natives: a phenomenon called "ecological facilitation." The yellow starthistle has been introduced into the West Coast and secretes the chemical compound 8-hydroxyquinoline from the roots. This compound harms native plants, thus allowing starthistle to increase its range as its chemicals wipe out native competitors.

Hassan Rasheed

Hazardous and Toxic Waste

A hazardous waste is waste that poses substantial or possible threats to public health or the environment. Worldwide, the United Nations Environmental Program (UNEP) has estimated the production of more than 400 million tons of hazardous waste universally each year by industrialized countries, with approximately 4 million tons shipped across international borders. The majority of transfers from industrialized countries to developing nations have to do with disposal because of the rising cost of disposing of hazardous waste in the home country. Disposal consists of storage either above or below ground, leaving a legacy for future generations. In the case of biological toxins, they are incinerated or released into air.

Exposure to hazardous waste is harmful to humans and wildlife. It is especially dangerous for young fledgling life forms. Fetuses, whether human or animal, are in a process of rapid development. The introduction of chemicals interferes with embryonic development, resulting in malfunctioning organs and limbs.

Hazardous waste stunts plant growth, much of which is useful to humans for consumption or manufacturing. In addition, the climination of plant life reduces the natural food supply for animals. Thus, hazardous waste exposure can potentially destroy an entire ecosystem.

The presence of hazardous waste in the environment is often the result of inadequate disposal. Improperly maintained landfills are a major issue. Although they appear to be isolated from contact with people or resources, they can still contaminate the surrounding environment. Waste in landfills sits for years or even generations. It emits gases that are both foul and toxic. It produces a liquid known as leachate, which can travel to water supplies such as rivers, lakes, and the ocean.

For example, gasoline is toxic and can cause environmental harm. It is the most popular fuel for small automobiles, and the transportation sector depends on it.

Plastics

Since their mass production began in the 1940s, plastics' wide range of unique properties has propelled them to an essential status in society. Next year, more than 300 million tons will be produced worldwide. The amount of plastics manufactured in the first 10 years of this century approaches the total produced in the entire last century, according to a report.

Evidence is mounting that the chemical building blocks that make plastics so versatile are the same components that harm people and the environment. Besides, their production and disposal contribute to an array of environmental problems. For example,

• Chemicals added to plastics are absorbed by the human body. Some of these compounds have been found to alter hormones or to have other possible effects on human health.

• Plastic debris, laced with chemicals and often ingested by marine animals, can injure or poison wildlife.

• Floating plastic waste, which can survive for thousands of years in water, serves as mini transportation devices for invasive species, disrupting habitats.

• Plastic buried deep in landfills can leach harmful chemicals that spread to groundwater.

• Approximately 4% of world oil production is used as a feedstock to make plastics, and a similar amount is consumed as energy in the process.

People are exposed to chemicals from plastics multiple times per day via air, dust, water, food, and use of consumer products.

For instance, phthalates serve as plasticizers in the manufacture of vinyl flooring and wall coverings, food packaging, and medical devices. Eight out of every 10 babies, and nearly all adults, have measurable levels of phthalates in their system.

In addition, bisphenol A (BPA), found in polycarbonate bottles and the linings of food and beverage cans, can leach into food and drinks. The CDC has reported that 93% of people have detectable levels of BPA in urine.

The report noted that the high exposure of premature infants in neonatal intensive care units to both BPA and phthalates is of "great concern."

Polybrominated diphenyl ethers (or PBDEs), which are flame-retardants added to polyurethane foam furniture cushions, mattresses, carpet pads, and automobile seats, are also widespread.

e-Waste

Roughly 40 million metric tons of electronic waste (e-waste) is produced globally each year, and ~13% of that waste is recycled mostly in developing countries. Approximately 9 million tons of this waste—discarded television sets, computers, cellphones, and other electronics—are produced by the European Union, according to the UNEP. This organization also notes that this estimate of waste is likely too low.

Informal recycling markets in China, India, Pakistan, Vietnam, and the Philippines handle anywhere from 50% to 80% of e-waste, often via shredding, burning, and dismantling the products in "backyards." Emissions from these recycling practices are damaging human health and the environment.

Developing countries with rapidly growing economies handle e-waste from developed countries, and from their own internal consumers. Currently, estimated 70% of e-waste handled in India is from other nations, but the UNEP estimates that between 2007 and 2020, domestic television e-waste will double, computer e-waste will increase fivefold, and cell phone waste 18-fold.

The black-market recycling practices magnify health risks. For example, primary and secondary exposure to toxic metals, such as lead, results mainly from open-air burning used to retrieve valuable components such as gold. Combustion from burning e-waste creates fine particulate matter, which is linked to pulmonary and cardiovascular diseases.

While the health implications of e-waste are difficult to isolate due to the unofficial working conditions, poverty, and poor sanitation, several studies in Guiyu, a city in southeastern China, provide relevant insights. Guiyu is known as the largest e-waste recycling site in the world, and the city's residents show substantial digestive, neurological, respiratory, and bone problems. For instance, 80% of Guiyu's children experience respiratory ailments, and are at an especially high risk of lead poisoning.

Residents of Guiyu are not the only ones at risk. Researchers such as Brett Robinson, a professor of soil and physical sciences at Lincoln University in New Zealand, warn that wind patterns in Southeast China disperse toxic particles released by open-air burning across the Pearl River Delta Region, home to 45 million people. This way, toxic chemicals from e-waste enter the "soil–crop–food pathway," one of the most significant routes of human exposure to heavy metals. These chemicals are not biodegradable: they persist in the environment for long periods, increasing exposure risk.

Global Warming

If you prepare a large capped glass flask containing ordinary air, with a green piece of filter paper inside, and expose it to sun rays, the temperature inside the flask will rise due to the various gases in the air and the color of the filter paper.

If you add pure carbon dioxide to the glass flask discussed above, the temperature inside the flask will rise more quickly. If you add pure methane, you will find that the warming effect is 21 times that of carbon dioxide. Similarly, if you add pure nitrous oxide, the warming effect is 330 times that of carbon dioxide.

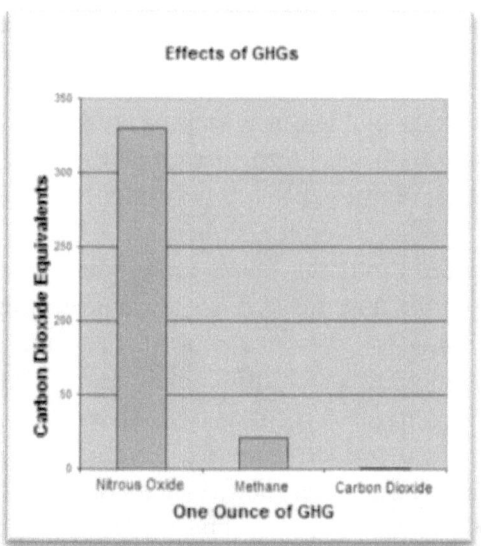

The effects of different greenhouse gases on temperature under controlled conditions in the laboratory

Today we are at the early stages of global warming and its effects, but there are signs that we are on the road to major changes in weather and ocean patterns.

Indeed, the water vapor content of the air is rising with an average increase in water vapor of .04 parts per million per year over Boulder, Colorado.

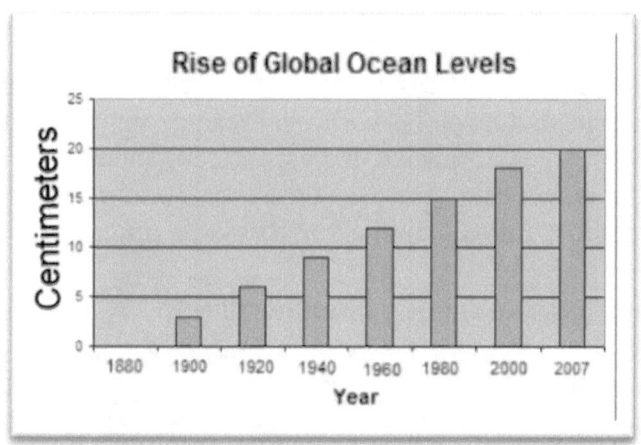

Actual sea level changes over the last one hundred years

Indeed, lower-latitude glaciers are melting away to the point where only 50% of them remain on Earth since the year 1900. There are many agricultural communities that are dependent on the annual melting of glacier waters and their replenishment.

In addition, the ocean and sea levels have been rising at the rate of an inch every 10 years, and this rise is now accelerating.

In the United States of America, forest fires are burning larger sections of land and are reaching higher up the mountains where there was once snow to slow them down. These fires are also burning hotter, and trees that depend on fires, e.g., the Ponderosa pines, are dying from the extreme heat the fires generate.

Nitrogen oxides (gases) perform an important function in atmospheric chemistry. They are emitted into the atmosphere naturally, mainly as a result of microbial activities in soils and lightning discharges. Currently and predominantly, the emissions occur as a result of human activities (such as combustion of fossil fuels, biomass burning, and the use of fertilizers). Long-term trends of their concentration in the atmosphere are not documented adequately. Nevertheless, reconstructed emission inventories suggest that large increases have occurred throughout this century. Exposure to nitrogen oxides has direct adverse effects on humans, animals, and plants. Nitrogen oxides also contribute to the global environmental problems facing our planet (i.e., excessive global warming, depletion of the ozone layer, and acid rain).

J. Hansen and colleagues report that melting of the ice sheets south and north of the globe will significantly raise ocean levels, cause severe weather storms and the hurdling of boulders into the air. These problems will reinforce our current march to global warming, which will reach a critical point in 10 years or so instead of hundreds.

Water Depletion and Contamination

Groundwater
Groundwater is a valuable resource throughout the world. Where surface water, such as lakes and rivers, is scarce or inaccessible, groundwater supplies many of the hydrologic needs of people everywhere. Sustained groundwater pumping as the result of

overpopulation causes groundwater depletion and is a key issue associated with groundwater use. Many areas of the world are experiencing groundwater depletion.

Pumping groundwater faster than its recharge rate can have some negative consequences for the environment and the people who make use of the water: Lowering of the water table increases the cost of drilling and pumping of water. In addition, the reduction in water amounts in streams and lakes is partly due to groundwater depletion and lowering of the water table.

Land subsidence occurs when water is removed from the soil. This means that the soil collapses and compacts rendering it useless to agriculture or wildlife.

Deterioration of water quality in fresh groundwater supplies is caused by saltwater intrusion from oceans, seas, very deep groundwater sources, and water below oceans that is saline. (USGS Fact Sheet and USGS Circular 1186) Water quality is also affected by mining, drilling for oil, and by toxic waste dumping where heavy metals and crude oil residues seep into ground water systems.

Lake and River Water

With the increased demand for fresh water in our growing populations, we are fast approaching a point where there will not be enough water to go around. The sources for lake and river water are groundwater in the form of springs, rain, glaciers, and mountain snow. With the advent of global warming and climate change, we can see that snow and glaciers are disappearing at an alarming rate as discussed earlier. This change affects the year-round supply of groundwater, lakes, and rivers.

With global warming, the temperatures of lake and river waters are gradually, increasing causing stress to aquatic fauna and flora. This stress can lead to the death or extinction of less hardy species and perhaps to introduction of species more tolerant of warmer waters in these waterways. In turn, these phenomena affect the food supply of many terrestrial species including humans that depend on these waters for food.

Cities

Before the advent of human technology, nature had figured out how to make the population stable. This was done by having the death and birth rates more or less equal. Since then, these rules have been ignored, and now the birth rate is higher than the death rate, resulting in an explosion in the human population from 2 billion a hundred or so years ago to 7.5 billion today.

With growing populations concentrated in cities and their suburbs, communicable diseases have flourished while domestic and industrial accidents made their debut. Garbage and human fecal waste had to be rid of. Garbage is typically buried underground where it leaks toxins into the groundwater systems and into air or is taken far out to sea to be dumped. Human fecal matter is sometimes treated and returned to the Earth, but most of the time, it is dumped in rivers or into the sea.

Antibiotics and Superbugs

Antibiotic resistance is a serious and growing problem in today's health care and has become one of the eminent public health concerns of this century. In the simplest cases, microorganisms acquire resistance via mutations that make them insensitive to common antibiotics, thereby requiring the use of less common secondary antibiotics. Primary antibiotics are preferred because they have several advantages that include safety, availability, and low cost. Secondary antibiotics are broader in spectrum, pose higher risks, and may be more expensive or locally unavailable. In the case of some multiple-drug-resistant pathogens, resistance to secondary and even tertiary antibiotics has been demonstrated, as seen in pathogens *Staphylococcus aureus* and *Pseudomonas aeruginosa* (which possess strong intrinsic resistance).

The major cause of widespread antibiotic resistance is the overuse of antibiotics in agriculture and in human health care. It is becoming harder and harder to fight infections caused by these drug-resistant pathogens, and their increase in number and prevalence is outpacing society's ability to fight them.

With the use of antibiotics, we have been killing off whole biologic communities in our soils, waterways, and putting ourselves in danger.

Colonization of the world

As the accumulation of wealth became institutionalized by the technological humans and reached a fever pitch, they embarked on the great expansionary conquest of the rest of the world not realizing that they were regressing populations that had achieved a stable equilibrium with their surrounding biota. Those who had achieved a balance with nature and were dancing the cyclical dance were doomed to a life of slavery serving the technological revolution and those who proclaimed own superiority.

Genetic Engineering

This technology is interfering greatly with natural selection. For example, there is genetically modified wheat that once planted becomes sterile. The way in which this phenomenon interferes with natural selection is that natural plant pollen is used up pollenating sterile plants, and the sterile pollen of the sterile plants pollenates the natural plants hindering their natural ability to multiply.

In addition, humans are sequestering more and more land to be cultivated for these sterile plants, thereby further interfering with nature's way of doing business.

Globalization

The promotion of free trade around the world and the spread of capitalistic economic systems have increased the spread of the above-mentioned dark events around the world.

Summary

The human bloom has not been kind to the Earth. Each of the preceding examples reveals how destructive we can be on the Earth's habitats doing ordinary things like going to work in the morning or taking a vacation.

It is obvious that we have not reached the evolutionarily mature energy-efficient stage. For example, a pride of lions does not finish off a herd of gazelle and move on. The pride of lions stake out a territory and consume only a part of the gazelle herd leaving plenty of them to breed and produce naturally more food for the pride. This approach ensures efficiency in the use of the sun's energy.

The pace of technological development fueled by abundant fossil fuel energy has outpaced evolution's ability to insert one or more genes into the human genome that would mitigate the environmental problems posed by humans on Earth and in Earth's biota. This is because human genetics takes anywhere from 100,000 to 500,000 years for any viable changes to be expressed, while the changes on Earth that are leading us to an inevitable mass extinction event are only 30 to 100 years away(See Limits to Growth in reading list).

A question must be asked at this time. "Is humanity's brain, with its accelerated learning abilities, able to replace its genetically controlled underpinnings and save the Earth?" In other words, can we think our way out of our death wish and achieve an efficient and mature way to use energy?

Rectification

With the highest so-called standards of living in the societies of the developed world, not all was sugar and spice to some humans, and they started to sound the alarm. Hopes were high that something can be done about the damage humankind has caused to its precious planet.

Sustainability

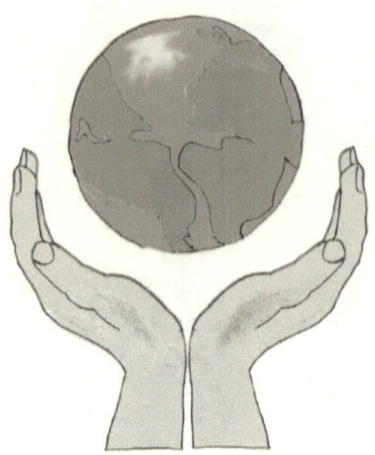

The idea of sustainability stems from the concept of sustainable development, which became a catchphrase at the World's first Earth Summit in Rio in 1992. Since then, there have been many variations and extensions on this basic definition. The following are but a few:

"Sustainability means development that meets the needs of the present without compromising the ability of future generations to meet their own needs"; Bruntland Report for the World Commission on Environment and Development (1992).

"A process of change in which the exploitation of resources, the direction of investments, the orientation of technological development and institutional change are all in harmony and enhance both current and future potential to meet human needs and aspirations"; The World Commission on Environment and Development.

"Sustainable development is a dynamic process which enables people to realize their potential and improve their quality of life in ways which simultaneously protect and enhance the earth's life support systems"; Forum for the Future.

"In essence sustainable development is about five key principles: quality of life; fairness and equity; participation and partnership; care for our environment and respect for ecological constraints – recognizing there are 'environmental limits'; and thought for the future and the precautionary principle"; from Making London Work by Forum for the Future's Sustainable Wealth London project.

"The environment must be protected... to preserve essential ecosystem functions and to provide for the wellbeing of future generations; environmental and economic policy must be integrated; the goal of policy should be an improvement in the overall quality of life, not just income growth; poverty must be ended and resources distributed more equally; and all sections of society must be involved in decision making"; the Real World Coalition 1996, a definition based on the work of the World Commission on Environment and Development.

"We cannot just add sustainable development to our current list of things to do but must learn to integrate the concepts into everything that we do"; the Dorset Education for Sustainability Network.

"A sustainable future is one in which a healthy environment, economic prosperity and social justice are pursued simultaneously to ensure the well-being and quality of life of present and future generations. Education is crucial to attaining that future"; Learning for a Sustainable Future – Teacher Centre.

"The first and perhaps most difficult problem, one that seldom gets addressed, is the time frame...Is a sustainable society one that endures for a decade, a human lifetime, or a thousand years?" (The Shaky Ground of Sustainable Development Donald Worster in Global Ecology 1993.)

"The overall goal of the UN Decade of Education for Sustainable Development (DESD) is to integrate the principles, values and practices of sustainable development into all aspects of education and learning. This educational effort will encourage changes in behavior that will create a more sustainable future in terms of environmental integrity, economic viability and a just society for present and future generations"; UN Decade of Education for Sustainable Development 2005–2014.

Many argue that sustainability has been hijacked and twisted to suit governmental and business interests that really want to continue with the way things are. For example, there is no quantifiable target of what the "sustainable future" looks like; there are no quantifiable parameters with which to measure any type of progress toward sustainability. In summary, sustainability is only a feel-good state of mind for human individuals, corporate leaders, and governments in the current state of our planet.

Conservation Biology

Conservation biology is tied closely to ecology in researching the dispersal, migration, demographics, effective population size, inbreeding depression, and minimal population viability of rare or endangered species. To better understand the restoration ecology of native plant and animal communities, a conservation biologist closely studies both their inclusive and exclusive habitats that are affected by a wide range of benign and adverse factors. Conservation biology is concerned with phenomena that affect the maintenance, loss, and restoration of biodiversity and the science of sustaining evolutionary processes that engender genetic, population, species, and ecosystem diversity. The concern stems from estimates suggesting that up to 50% of all species on the planet will disappear within the next 50 years; this situation has contributed to poverty, starvation, and will reset the course of evolution on this planet.

Conservation biologists research and educate on the trends and process of biodiversity loss, species extinctions, and the negative effect these are having on our abilities to sustain the well-being of human society. Conservation biologists work in the field and office, in government, universities, nonprofit organizations, and industry. They are funded to research, monitor, and catalog every angle of the Earth and its relation to society. The topics are diverse because this is an interdisciplinary network with professional alliances in biological and social sciences. Those dedicated to the cause and profession advocate a global response to the current biodiversity crisis based on morals, ethics, and a scientific reason. Organizations and citizens are responding to the biodiversity crisis via conservation action plans that direct research, monitoring, and education programs that deal with concerns from a local to global scale.

Environmental Protection Agencies

Environmental protection agencies or organizations have cropped up all around the world. The United States Environmental Protection Agency (EPA or sometimes USEPA) is an agency of the U.S.A. federal government which was created for the purpose of protecting human health and the environment by writing and enforcing regulations based on laws passed by Congress. The EPA was proposed by President Richard Nixon and began operation on December 2, 1970, after Nixon signed an executive order. The order establishing the EPA was ratified by committee hearings in the House and Senate. The agency is led by the Administrator, who is appointed by President and approved by Congress. The current administrator is Gina McCarthy. The EPA is not a cabinet department, but the Administrator is normally given the cabinet rank. Among EPA programs are the following:

(1)   Energy Star
In 1992, the EPA launched the Energy Star program, a voluntary program that fosters energy efficiency.

(2)   Pesticides
EPA administers the Federal Insecticide, Fungicide, and Rodenticide Act and registers all pesticides legally sold in the United States.

(3)   Environmental Impact Statement Review
The EPA is responsible for reviewing Environmental Impact Statements of other federal agencies' projects, under the National Environmental Policy Act (NEPA).

(4)   Safer Detergents Stewardship Initiative
Through the Safer Detergents Stewardship Initiative, EPA's Design for the Environment recognizes environmental leaders who voluntarily commit to the use of safer surfactants. The latter are the surfactants that break down quickly into nonpolluting compounds

and help protect aquatic life in both fresh and salt water. Nonylphenol ethoxylates, commonly referred to as NPEs, are an example of a surfactant class that does not meet the definition of a safer surfactant.

(5)   EPA Safer Choice

The EPA Safer Choice label, previously known as the Design for the Environment label, helps consumers and commercial buyers identify and select products with safer chemical ingredients, without sacrificing quality or performance. When a product has the Safer Choice label, it means that every intentionally added ingredient in the product has been evaluated by EPA scientists. Only the safest possible functional ingredients are allowed in products with the Safer Choice label.

(6)   Fuel economy

Manufacturers selling automobiles in the United States are required to provide EPA with fuel economy test results for their vehicles, and the manufacturers are not allowed to provide results from alternate sources. The fuel economy is calculated using the emissions data collected during two of the vehicle's Clean Air Act certification tests by measuring the total amount of carbon captured from exhaust fumes during the tests.

(7)   Air quality

The Air Quality Modeling Group is in the EPA's Office of Air and Radiation and provides leadership and direction on the full range of air quality models, air pollution dispersion models, and other mathematical simulation techniques used for assessing pollution control strategies and the effects of air pollution sources.

(8)   Oil pollution

The Spill Prevention, Control, and Countermeasure Rule apply to all facilities that store, handle, process, gather, transfer, store, refine, distribute, use, or consume oil or oil products. This definition includes petroleum and nonpetroleum oils as well as animal fats, oils, and greases; fish and marine mammal oils; and vegetable oils, (including oils from seeds, nuts, fruits, and kernels).

(9)   EPA WaterSense

WaterSense is an EPA program designed to encourage water efficiency in the United States through the use of a special label on consumer products. It was launched in June 2006. Products include high-efficiency toilets (HETs), bathroom sink faucets (and accessories), and irrigation equipment. WaterSense is a voluntary program, with the EPA developing specifications for water-efficient products through a public process and product testing by independent laboratories.

(10)   Drinking water

The EPA ensures safe drinking water for the public, by setting standards for more than 160,000 public water systems nationwide. The EPA oversees states, local governments, and water suppliers to enforce the standards under the Safe Drinking Water Act. The program includes regulation of injection wells in order to protect underground sources of drinking water. Selected readings of amounts of certain contaminants in drinking water, precipitation, and surface water, in addition to milk and air, are reported on EPA's Rad Net web site.

(As of this publication many of the EPA regulations are said to be rolled back erasing the progress mentioned here.)

Development of Metrics: Ecological Footprint

There have been many attempts to try to give teeth to the concept of sustainability. One of the early ones is the concept of carrying capacity of land. For example, this concept is supposed to describe how many humans can be supported on 10 acres of forest land.

Another concept is that of an ecosystem. This notion is intended to describe the relations or links between the air, soil, water, plants, and animals of a forest or a desert and in turn provide an understanding of how humans can be supported. Ecosystems are complex and hard to measure and describe fully.

Both these concepts have fallen short of giving a true picture of what it will take to support a human. With respect to the human carrying capacity of a plot of land, it was not useful because of our global economy. For example, Egypt imports 50% of its food from other countries; therefore, it is impossible to measure the carrying capacity of an acre of land in Egypt.

A recent measure reflecting the effects of humans on Earth's natural dance of rejuvenation is the ecological footprint. The latter is a standardized measure of human need for Earth's ecosystems. It measures the human demand for natural resources that are contrasted with the Earth's ability to regenerate them. It represents the amount of biologically productive land and sea areas necessary to supply the resources humans consume and to assimilate the resulting waste if a population follows a given lifestyle.

The first academic publication about ecological footprints was authored by William Rees in 1992. The ecological footprint concept and calculation method were developed as the PhD dissertation of Mathis Wackernagel, under Rees' supervision at the University of British Columbia in Vancouver, Canada, from 1990 to 1994. In 2003, Jason Venetoulis, Carl Mas, Christopher Gaudet, Dahlia Chazan, and John Talberth developed Footprint 2, which offered a series of theoretical and methodological improvements to the standard footprint approach. The four primary improvements were as follows: they included the entire surface of the Earth in bio-capacity estimates, allocated space for other (i.e., nonhuman) species, updated the basis of equivalence factors from agricultural land to net primary productivity (NPP), and refined the carbon component of the footprint based on the latest global carbon models.

Global Ecological Footprints: Currently there is no established way to measure global ecological footprints, and any attempts to describe the capacity of an ecosystem as a single number are massive simplification of thousands of key renewable resources, which are not used or replenished at the same rate. Nevertheless, there has been some convergence of metrics and standards since 2006.

That said, City Ecological Footprints measurements are being attempted. There are two types of measurements. The first evaluates ecosystem displacement, which is defined as City Area minus the remaining green spaces. This is an area measure that does not include human or other biological activity. The second type is designed to quantify health of the surviving ecosystem. Specifically, it is supposed to quantify both the area and biological health of ecosystems surviving inside city areas such as nature reserves, parks, and other green spaces. City ecological footprints are being calculated and ranked with city ecological indexes.

The more detailed methods of calculating the ecological footprint are presented by Rees and Wackernagel in the book entitled *Our Ecological Footprint: Reducing Human Impact on the Earth.*

For 2007, humanity's total ecological footprint was estimated at 1.5 planet Earths; that is, humanity uses up ecological resources 1.5-fold more quickly than the Earth can renew them. We have been living off of the principal in our bank account as well as the interest. In other words, to get to the fruit, we have been cutting down its tree.

When we examine an undisturbed ecosystem, we find that each species takes only as much as can be regenerated. That means that each species has zero effect on the ecosystem and we can safely say that each species occupies an ideal zero ecological footprint.

Roadblocks

As the human bloom continues unabated, destroying every living cycle in its path, we need to ask why the human species with its intellectual prowess has not been able to produce a mature strategy and target for human existence and survival of the planet.

The Eternal Optimists

There are those of us who consider their cup perpetually full. They have every belief that faith or technology or something will always come to the rescue.

Denial

The root of denial is the lack of education and understanding of the complexities of the current situation facing us. Many who are enjoying their lifestyles are not concerned about what is going on with the environment, do not believe that it is all that serious, or have a problem believing that it is all going to end sooner rather than later.

This is especially true with seniors who consider their time on Earth limited. The younger crowds are too busy working to simply look up and see what is really going on. The optimists are always thinking that science and technology will simply solve any problems that come our way, for instance, by substituting one product for another or finding ways of cleaning up the mess we have created for ourselves.

Goals of Belief Systems

It is normal for humans to want to know the beginning, the middle, and end of things which include their story, their purpose,

74

ideals, what happens after death, and the story of their universe. These are told in a language that the cooperative group understands, taking images from the surrounding landscape and wisdom from its leaders.

Belief systems are a reflection of what is in the hearts of humans as they try to make sense of the world. Belief systems are what guide us through our daily thoughts and activities. They give us a sense of what is right and what is wrong with respect to family, community, and strangers. They define who we are as an individual or as a nation.

Belief systems (such as the current amalgam of major religions and philosophies of life and government) rarely if ever mention the importance of maintaining or replenishing what we use of the Earth itself. One only has to look out the window and observe. Most of the Earth has been denuded of its native species, whether plant of animal, and replaced with domesticated ones. It is as if we have the right to do so and that we own this planet and can do with it whatever we please.

Goals of Governments

Governments take on the characteristics of its people or rulers and their belief systems. As with the command and control center of a cell or an animal, a modern nation has a set of laws and regulations (constitution) to be followed by all people in their pursuit of happiness. In a survey of 183 countries, most if not all defined in their constitutions in one way or another, their territories and resources, advocating the freedom of the individual to live and prosper within own territory using its resources.

Only ~50% of these countries have a constitution that mentions preservation of the environment or maintenance of the ecology within its territory. For instance, the following are statements from several constitutions in this survey:

African Nation of Namibia
1. The duty to investigate complaints concerning the overutilization of living natural resources, the irrational exploitation of non-renewable resources, the degradation and destruction of

ecosystems and failure to protect the beauty and character of Namibia.

2. Maintenance of ecosystems, essential ecological processes and biological diversity of Namibia and utilization of living natural resources on a sustainable basis for the benefit of all Namibians, both present and future; in particular, the Government shall provide measures against the dumping or recycling of foreign nuclear and toxic waste on Namibian territory.

Middle Eastern Nation of Egypt
Article 44: The Nile
The state commits to protecting the Nile River, maintaining Egypt's historic rights thereto, rationalizing and maximizing its benefits, not wasting its water or polluting it. The state commits to protecting its mineral water, to adopting methods appropriate to achieve water safety, and to supporting scientific research in this field.

Every citizen has the right to enjoy the Nile River. It is prohibited to encroach upon it or to harm the river environment. The state guarantees to remove encroachments thereon. The foregoing is regulated by law.

Article 45: Seas, beaches, lakes, waterways, mineral water and natural reserves
The state commits to protecting its seas, beaches, lakes, waterways, mineral water, and natural reserves.

It is prohibited to encroach upon, pollute, or use them in a manner that contradicts their nature. Every citizen has the right to enjoy them as regulated by law. The state also commits to the protection and development of green space in urban areas; the protection of plants, livestock, and fisheries; the protection of endangered species; and the prevention of cruelty to animals. All the foregoing takes place as regulated by law.

Article 46: Environment
Every individual has the right to live in a healthy, sound and balanced environment. Its protection is a national duty. The state is committed to taking the necessary measures to preserve it, avoid harming it, rationally use its natural resources to ensure that

sustainable development is achieved, and guarantee the rights of future generations thereto (22 Draft dated 2 December 2013 of the Constitution of the Arab Republic of Egypt).

The European Nation of Finland
Section 20 – Responsibility for the environment
Nature and its biodiversity, the environment and national heritage are the responsibility of everyone. The public authorities shall endeavor to guarantee for everyone the right to a healthy environment and for everyone the possibility to influence the decisions that concern their own living environment.

Constitution Summary

On average, if you add up all the constitutional texts of the 186 countries surveyed, the statements involving the preservation of the environment and its ecology represent pitiful 0.2% of the sentence count. Again, we see that an overwhelming majority of the laws on Earth today do not reflect the reality: a finite and cyclical nature of its biota. It is true that some nations have organizations that are actively trying to protect the environment like environmental protection agencies, but these have mostly been knee jerk reaction agencies trying to rectify massive damage that has already been done lacking specific goals to actually maintain a healthy ecosystem or eliminate humanity's damaging ecological footprint.

Goals of Corporations and Companies

As with belief systems and governments, businesses have a figurehead who orchestrates the business plan and mission statement or goals that have been written down for all involved to follow. The following is a small sample of mission statements of leading businesses around the world taken from their websites. Again, concern for well-being of the Earth, its environments, and a reduction in humanity's ecological footprint are lacking.

Bruegger

Bruegger bakes authentic New York-style bagels fresh from the oven all day. With the coffee freshly brewing and a full selection of breakfast and lunch sandwiches, wraps and Paninis, specialty soups and salads, our team creates a memorable experience that makes people say, "Let's go to Bruegger's!"

Nike

"To lead in corporate citizenship through proactive programs that reflect caring for the world family of Nike, our teammates, our consumers, and those who provide services to Nike."

Unilever

Unilever USA, the parent company of Lipton Tea, outlines its mission as working to create a better future every day helping people feel and look better and get more out of life with brands and services that are both good for them as well as others, inspiring people to take small actions that make big differences in the world, and developing new ways to double the business, as well as reduce the environmental impact of the business.

International Business Machines (IBM)

At IBM, we strive to lead in the invention, development and manufacture of the industry's most advanced information technologies, including computer systems, software, storage systems and   microelectronics.

We translate these advanced technologies into value for our customers through our professional solutions, services and consulting businesses worldwide.

Hertz

"Our mission is to be the most customer focused, cost efficient vehicle and equipment rental/leasing company in every market we serve. We will strengthen our leading worldwide positions through a shared value culture of employee and partner involvement by making strategic investments in our brand, people and products. The focus of everything we do will be continuously improving shareholder value."

Boeing
Leadership

We will be a world-class leader in every aspect of our business—in developing our team leadership skills at every level; in our management performance; in the way we design, build and support our products; and in our financial results.

Integrity

We will always take the high road by practicing the highest ethical standards and by honoring our commitments. We will take personal responsibility for our actions and treat everyone fairly and with trust and respect.

Quality

We will strive for continuous quality improvement in all that we do so that we will rank among the world's premier industrial firms in customer, employee and community satisfaction.

Customer satisfaction

Satisfied customers are essential to our success. We will achieve total customer satisfaction by understanding what the customer wants and delivering it flawlessly.

People working together

We recognize that our strength and our competitive advantage is—and always will be—people. We will continually learn, and share ideas and knowledge. We will encourage cooperative efforts at every level and across all activities in our company.

A diverse and involved team

We value the skills, strengths and perspectives of our diverse team. We will foster a participatory workplace that enables people to get involved in making decisions about their work that advance our common business objectives.

Good corporate citizenship

We will provide a safe workplace and protect the environment. We will promote the health and well-being of Boeing people and their families. We will work with our communities by volunteering and financially supporting education and other worthy causes.

Enhancing shareholder value

Our business must produce a profit, and we must generate superior returns on the assets entrusted to us by our shareholders. We will ensure our success by satisfying our customers and increasing shareholder value.

Core competencies

Detailed customer knowledge and focus

We will seek to understand, anticipate and be responsive to our customers' needs.

Large-scale systems integration

We will continuously develop, advance and protect the technical excellence that allows us to integrate effectively the systems we design and produce.

Lean enterprise

Our entire enterprise will be a lean operation, characterized by the efficient use of assets, high inventory turns, excellent supplier management, short cycle times, high quality and low transaction costs.

Strategies

Run healthy core businesses.

Leverage strengths into new products and services.

Open new frontiers.

People working together as a global enterprise for aerospace leadership.

Mission Statement Summary

Out of the six above-mentioned mission statements for leading companies on Earth, only one mentions the environment, and even in that one case, it is only to reduce and not eliminate the company's impact. Again, we find a lack of urgency in matters of the ecological stability of the planet. Perhaps this is because the bottom line of every business is its profitability compared to its competitors. Profitability is where the company takes more than what it gives out. This profit is essential for growing the company, and the growth is another goal that every for-profit organization tries to achieve. Perhaps there is still a belief that the resources on Earth are vast and endless.

Resistance to Change

As is often the case in the pursuit of sustainability or conservation, if the reason for a change is ambiguous, whether it is about costs, equipment, or jobs, it can trigger negative reactions among people.

Another reason people resist change is when they have not been consulted or involved or when the change is forced upon them. Change also carries a threat to established patterns of working and social relations among people, or it has not been communicated well to the people involved.

The third reason change is a threat is when the benefits and rewards for making the change are not seen as worth the trouble involved. In addition, when the change threatens jobs, power, or status in an organization, it is doomed to fail.

When change affects people in power more than it does the regular Joe, these people put up the highest resistance.

Colossus Economies

Sketch of a P&H Mine Pro Shovel

Please visit Appendix 1, which discusses the birth of economic thought, where a question is posed: "Where do you think this dynamic endless loop of supply and demand has led us?" The answer is that it has led us to what can be described by one phrase, "*Economies of Scale*." In other words and generally speaking, it has been discovered that the greater the scale of the manufacturer or producer, the lower is the cost per unit output.

An old example may clarify the point. Adam Smith published a book in 1776 on economics entitled *The Wealth of Nations* describing pin making. He explained that one pin maker with little experience can make one to ten pins a day. Then, he described a group of 18 people all involved in the making of pins where they divided the process into 18 steps, where one person performed a single step in the process, the result being that this group was able to manufacture 48,000 pins a day.

As a result, the cost of making a pin was far lower when the company of pin makers was larger and they adopted the methods described above that are also known as *The Division of Labor*. A modern example is large jumbo jets that carry 400 to 600 passengers at a time. To an airline, the cost per seat or mile is much lower in comparison with an aircraft that carries 100 passengers, and the reduced cost is passed on to the consumer.

When several companies are competing to provide the lowest cost item to the public who are growing in numbers and who want cheaper goods or services with equal or better quality, these companies tend to go for economies of scale, and generally speaking, building or purchasing larger and larger equipment, factories, or products with the end results discussed in the section on *The Assault* and its undesirable outcomes.

The sad part of economies of scale is that we are locked into them, and it is hard to go back. Reversing economies of scale is not like downsizing a company to reduce cost but like raising the cost of living considerably. It just does not make sense to the average person.

It is a good feeling when a person is able to make a cheaper widget or purchase one. It is natural after all. Nonetheless, this natural feeling is carried forward from a time when higher energy efficiency was the norm. The value of fossil fuels dictated by monetary systems is interfering with our ability to devise ways to truly sustain or conserve a lifestyle suitable for the Earth.

The Dichotomy

On the one hand, we have the companies and corporations that are competing to make less expensive widgets and the consumers trying to find reasonably priced ones. As economies of scale come into play, we find companies moving to locations where they can build mega factories and where the public knows them only by brand name, with human populations increasingly moving to larger cities.

This separation of the producers from the consumers creates a vacuum of information requiring an additional layer of government watchdogs to make sure that producers are behaving properly in developed countries. This state of affairs becomes more complex when the producers go global and in most instances face no watchdogs at all.

This vacuum of information makes it exceedingly difficult for any decisions on the well-being of the planet to be carried out. On the one hand, we have the developed country's consumers enjoying less expensive widgets and the third world countries only glad to fulfill this need by accepting the mega factory's promise of a better life for its people.

It is true that the United Nations has succeeded (after many years) in reaching a consensus on carbon emission controls that have yet to be instated and will take many more years to implement, but carbon emissions are only a fraction of the problem as we saw in the section on *The Assault*. This situation highlights the problem in the approach to solving the catastrophic environmental problems we face. There is no consensus on the root cause of our environments problems, and instead we are trying to tackle one symptom at a time out of a large multitude.

Tolerance

I lived in the city of Springfield, Oregon, back in the 1970s. It is a small city in comparison to Portland, but that was an advantage because it was at the far south end of the Willamette Valley where the coastal and Cascade Mountains met. It was a beautiful scene in timber country.

Unfortunately, when the wind was coming from the east, the stench of rotten eggs would overcome me. Yet no one complained. When I asked why, I was told by my grandfather that it was the smell of money. You see, the paper plant then was the largest employer.

Humans have a great ability to accept an inconvenience if in the end they benefit. It is only when this nuisance is coupled with the knowledge of personal danger that something is done. And so it was with the smog that filled the air in the 1960s and 70s when large cities started to be affected by the exhaust of autos and power plants. It wasn't until the dangers of lung diseases from such polluted air became known that the danger signs went up causing the introduction of emission controls and transplanting of coal or oil electric generator plants away from large cities.

When we think of the benefits of using buses for transportation within cities, that is a good thing. But when we think of competition in making cheaper buses leading to economies of scale, with its assault on the environment, that is a bad thing. We get confused by the need to hop on the do-good bandwagon only to find out that it also does us and the environment harm.

We are therefore unable to think of a conservation target in favor of the environment because it will eventually bite us where it hurts: in the pocket book. Some of us try to think of targets only to find out they eventually will not work because of our dependence on the colossal economy. For example, I have read books on how to "leave civilization" and live off of one to three acres of land by farming it one way or another. Unfortunately, at the end of the day, we need to go back to our homes (that take forest land to grow) with all their modern-day conveniences (steel pots and pans that require mining of iron and processing in mega factories).

Environmental Politics

It is a sorry state when the speaker of the United States of America's House of Representatives says that environmental scientists are using pseudo-science to make their predictions of the calamities facing our nation and the globe. Our scientists are the brains of our country, and all they are doing is measuring and reporting on what they see. They don't have an ax to grind.

It is therefore exceedingly difficult for our nation to make any kind of informed decisions on the environment with such attitudes of people wielding so much power.

Apathy

Some have realized that the damage done to the biosphere can't be reversed, and consequently they give up and voice the opinion that any suggestions to change our ways of life are fruitless, will not happen, or are irrational. They claim that there have been several catastrophic events on Earth in the past 4 billion years, so why should we worry about this one. After all, those living today will not be around when it happens.

Intentions and Birth Control

There have been attempts to gain control of the human population in the 1960s and 1970s. This was mostly done in third world counties that were thought to have an exploding population. The attempts incorporated birth control options that included sterilization.

This effort backfired. Many third world countries objected to mandatory birth control especially sterilization on the grounds that it discriminated between rich and poor nations. Poor nations felt that rich nations were imposing their will upon the poor.

That is not to say that sterilization was not attempted in the United States of America. It suffered a similar fate.

Summary

The roadblocks that are keeping humans from formulating a global goal or target for our societies that would take us all out of our current tragic situation are many, and so the human bloom continues unabated undermining the cyclical nature of life on Earth. Whenever a voice is raised on behalf of the Earth regarding what action needs to be taken to formulate an all-inclusive target for all to follow, there is a dozen voices from individuals, companies, and governments, who support our current lifestyles, cry foul, , or "it is pseudo-science", or "it can't be done" or are silent.

There is also ignorance of what the basic problems are. For example, very few realize that the nature of life on Earth consists of energy efficient cycles that can be studied, measured, and quantified. Without this knowledge, a unified global or local societal target can never be formulated.

Thus, a unified target to save the biota (of which we are an integral part) on Earth has never been achieved. The next part of this publication explains a fair and equitable societal target to aim for that takes into consideration the needs of humans as well as the needs of the biota on Earth.

Target

Mother Earth has been dancing in the light of the Sun, is billions of years old and through it all has learned a thing or two, by building self-correcting behaviors into the biota to maintain the dance of life giving pleasures to the participants and hardships to those who veer away.

After the great assault of the humans and the spreading of their ways of assuming that every setback is a challenge to their notions of grandeur instead of a lesson in how to become wise, they stopped listening to the subtle tempo of the biotic music that was being played and tried to overcome it. It did not matter what catastrophe swept through their overcrowded communities: no locust swarms or crop failures or famine or plague stood in their way. They had to overcome nature as if it was an enemy and not their friend that brought them news of the right path to a peaceful and rich life of play, fun, rhythm, and song.

Humans had expansionistic technologies on their side boosting their egos with thoughts of splendor and magnificence thinking that these technologies contained a kernel of wisdom on which the world ran; that nature was so-called primitive and not of human form, to be despised and trampled upon. How wrong they were and are!

Despite nature's repeated attempts to sound an alarm, humans continued to poison the streams of advice flowing though the lands now filled with toxic hopes of reaching asteroids for the minerals that will become exhausted here on Earth. So, now human veins are carrying poisoned blood, and human thoughts are concentrated on where else to dump the outcomes of their actions so that nothing can be seen. Technology has robbed them of a future free of foolishness.

Humans once had friends with which they played and danced to the music of blissful existence, but since they chose to cease brotherly relations with the rest of life forms, they have no one to pull them back into the fold of mutual love and respect. They stand alone in their quests with no one to help them out when they fall. They have cut all ties with their family and have no one to lean on when the inevitable comes.

The humans had to toil with that which can't be toiled with. It will not be until the last individual is born that they will realize that they should have listened to the Mother, the Father, and the rest of their kin.

Return to Nature

Be advised that this section and those that follow are not for the faint-hearted. They contain language and ideas that are contrary to most of current sensibilities. They present facts about our current and future lifestyles. They discuss a unified global lifestyle to save the biota on Earth and much of what our future is to be like if we are to survive but does not propose how we get there. That task is up to you.

Zero Ecological Footprint

This means that in a biological cycle what is taken locally is returned in full locally.

For example, if we take the diagram above, the starting point is the bush, which converts sunlight into bioenergy. The bush also takes up minerals from the Earth. It is then partially eaten by the goat that is partially eaten by the human, and so on. The feces of the human are then partially consumed by the fly's offspring followed by the bird eating the flies and so on around the cycle. The table below shows a simulation of the cycling of energy and minerals. It implies that the bush has 100 units of energy (of which 50% are consumed by the goat) and 10 units of minerals (50% of which are also consumed by the goat). In addition, only 80% of the consumed energy and minerals are metabolized, and the rest are defecated.

Consumption			Used		Defecated		Totals	
Species	Consumed	Metabolized	Energy	Mineral	Energy	Mineral	Energy	Mineral
Goat	50%	80%	40.00	4.00	10.00	1.00	50.00	5.00
Human	50%	80%	16.00	1.60	4.00	0.40	20.00	2.00
Fly	50%	80%	1.60	0.16	0.40	0.04	2.00	0.20
Bird	50%	80%	0.64	0.06	0.16	0.02	0.80	0.08
Amoeba	50%	80%	0.06	0.01	0.02	0.00	0.08	0.01
Bacteria	50%	80%	0.03	0.00	0.01	0.00	0.03	0.00
Bio-degradable			21.67	2.17	5.42	0.54	27.09	2.71

The balance sheet for the above cycle

The last column in the balance sheet above represents the total amounts consumed (metabolized and defecated) and not used (biodegradable material that becomes a part of the ground humus). Of course, the balance sheet above is a simplification of the processes in the cycle. For example, it doesn't show catabolism, but you can build your own table that includes it.

The take-away insight from the diagramed cycle and its balance sheet above is that energy and minerals are conserved. Of course, you must realize that energy metabolized can serve many functions not demonstrated here. It can be used for movement, warmth, and perspiration as well as building muscle or hair. In addition, readers must realize that all energy and minerals are conserved locally.

The Target Human Population Count

The truth of the matter is that humans evolved into a diverse biota, and their genetic makeup had to adapt. Once one thinks along the lines of humans' being just one aspect of Earth's diversity and that all life is interlocked in the cyclical dance of existence, it becomes clear that we all depend on each other and that breaking of even one dancing cycle will ultimately unravel the whole web of existence.

Therefore, the question of the target should consider humans not as the top of the food chain but as an equal and integral part of life on planet Earth. Notions like "humans have superior intelligence" have continuously crept into the literature as an expression of our naturally blind drive to acquire more resources than our perceived competitors, enemies, or anything that stands in our way. It must be emphasized that humans are a link no better or worse than any other link, be it a clump of mud or another species in the web of life. As for human intelligence, it has been subservient to the expansionistic life stage present in all species (be they amoebas, birds, or dinosaurs): to steal, hoard, and defeat the enemy to provide for themselves and their young.

This expansionistic life stage worked well for humans at their early developmental stage when the Earth seemed vast and endless, and the opportunistic spirit was alive and well. Today, it is neither vast nor endless, and we are fast approaching critical limits of the current expression of our expansionistic life stage in a finite world. We must find a way to move on with our development to the mature energy-efficient stage and at the same time to reverse our paths from those that created these critical limits and to work to eliminate them. This means that we must reverse course and build up the principal in our bank accounts. We must share the interest with all other species. Instead of cutting down the tree to get to the fruit, we should wait for the fruit to ripen and perhaps fall to the ground before we eat it.

Our aim should be to have no impact on the Earth. Our aim is to have zero ecological footprint like that of the lioness we covered previously. As it turns out, the ideal human population count should be between 10 and 50 million individuals (see Appendix 3). This estimate is based on available land and sunlight. Upcoming chapters will describe the lifestyle and other conditions needed to support this population count at zero ecological footprint.

Hassan Rasheed

Human Niche

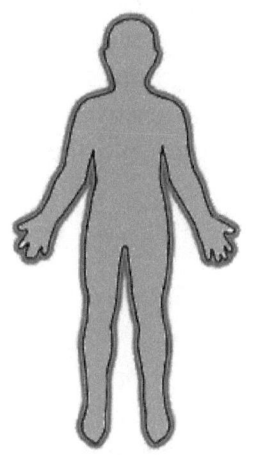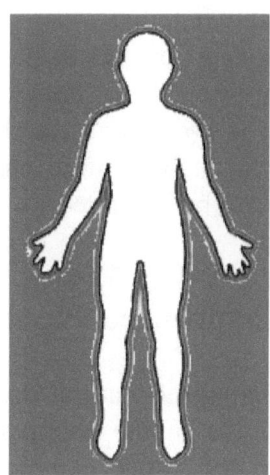

We need to get in touch with our original selves out from the shadows of our artificial modern lives. We will never know for sure what the original human niche was like because that bridge has been crossed and was washed away maybe 10,000 years ago. We can't stop wondering, though, what it was like to fit the environment like a hand fits a finely tailor-made glove because it was in that niche that we were the most efficient. It was in that niche that we enjoyed an inborn freedom as simple as plucking a wild apple from its tree and not worrying where the next one will come from. That niche is where natural zero ecological footprint was the norm.

You might ask, "Has the human condition changed with respect to its original niche?" The answer is yes and no. As far as our genetics, it probably has not changed that much because of the length of time it takes for a viable genetic trait to appear on its own; something like 100,000 years. But as far as the physical environment is concerned, it has changed, as discussed in the section on *Darkness*. In addition, our behaviors have changed because today our lives are wrapped up in large societies that are different from those of the hunter gatherer people's small groups, as you will see.

There are clues to what it must have been like. We can look at our closest relatives who are alive today or who perished not long ago and imagine. From there, we can look at how different we are from them and what that must have meant and therefore see the difference between our and their niches. We will first look at the chimpanzee followed by the hunter gatherer groups from the past and the present. It is hoped that the reader will discover semblances of his or her physical, behavioral characteristics and the environment for which they are better suited.

The Chimpanzee: Physical Characteristics

(1) Chimpanzee Natural Habitat

The habitat of the chimpanzee is one that is diverse. The rainforests are the most common areas where they are found. Nonetheless, they also live in the savannahs and in mountain forest areas. They are adaptable to a variety of environments. This is why their habitat continues to be one that is vast and diversified.

In a rainforest, they are able to find plenty of trees for hiding and for sleeping. It also makes it easier for them to gain access to fruits, plants, leaves, and seeds as well as shaded areas. In savannahs, chimpanzees are in open areas that have scattered woodlands.

A lower number of chimpanzees are found in mountainous regions because of the climate. It is also possible that some moved there in search of food, once their habitat had been taken from them by humans or other species.

Chimpanzees do need water, and therefore they will be found in areas where there is enough food and a good supply of water. This supply can come from a stream of major rivers.

A chimpanzee troop usually has a home range. The size of this area can vary from ~5 to 400 km^2. It often depends on the regenerative properties of the area and the number of chimpanzees in the troop. When the home range is large, the group will often split up and forage into different areas of the range.

The emigrational pattern through the home range is the one that is complex. It ensures that chimpanzees can get to certain areas where seasonal fruits grow just in time. The movement prevents them from depleting any single area of the food sources that are available there. In fact, their methods of eating often ensure that new vegetation can grow back for future use.

(2) The Fossil Record

New research demonstrates that hominins (early human species) in what is today northern Africa lived equally well in a relatively warm and dry climate 3.4 million years ago and in a much cooler climate with significantly more rainfall and forest growth slightly later. *Australopithecus afarensis* adapted to these dramatic environmental changes without the benefit of an enlarged brain, which aided later hominins in adapting to their environments.

These findings contribute to an ongoing debate about whether hominins of the Pliocene era preferred settings that were open and arid or wooded and moist, or whether they could adapt well to diverse environments. A lack of data on changes in past ecosystems to compare with hominin fossil data has hampered the inquiry.

Researchers analyzed fossil pollen located in stratified rock formations near Hadar, Ethiopia. Using these samples, the researchers identified three persistent plant communities: steppe and tropical and temperate forests containing water-conserving plants. A fourth plant community, forests containing plants that grow in cooler and wetter climates, appears and disappears in the pollen record. The presence of this fourth community corresponds to climate records of cooler and wetter periods.

These early humans had the ability to adapt to environmental changes. They could live in arid grasslands and forested surroundings as well.

Other researchers reconstructed likely responses of human ancestors to the climate of the past 5 million years using genetic modeling techniques. When results were mapped against the timeline of human evolution, the researchers found that key adaptive events coincided with periods of high variability in recorded temperatures.

The study confirmed that a major human adaptive radiation— a pattern whereby the number of coexisting species increases rapidly before crashing again to near previous levels—coincided with an extended period of climatic fluctuation. After the onset of high climatic variability ~2.7 million years ago, a number of new human species appeared in the fossil record, with most disappearing by 1.5

million years ago. The first stone tools appear at ~3.2 million years ago and doubtless assisted some of these species in responding to the rapidly changing climatic conditions.

By 1.5 million years ago, there was a single human ancestor left: *Homo erectus*. The key to the survival of *H. erectus* appears to be its behavioral flexibility: it is the most geographically widespread species of the period, and endured for over 1.5 million years. While other species may have specialized in environments that subsequently disappeared—causing their extinction—*H. erectus* appears to have been a generalist, able to deal with many climatic and environmental conditions.

Variability selection suggests that evolution, when faced with rapid climatic fluctuation, responded to the range of habitats encountered rather than to each individual habitat in turn; the timeline of variability selection suggests that *H. erectus* could be a product of exactly this process.

Linking climatic fluctuation to the evolutionary process has implications for the current global climate change debate. What we see in many areas of the world today is in fact an increased annual range of temperatures and conditions; this means in particular that third world human populations, many living in what are already marginal environments, will face ever more difficult situations. The current pattern of human-induced climate change is unlike anything we have seen before and is disproportionately affecting areas whose inhabitants do not have the technology required to deal with this change.

Physical Evidence Summary

Judging by the physical evidence, it appears that our early ancestors had several niches to deal with because of the constantly changing environment. They became jacks-of-many-trades. This transition required changes to their morphology such as a larger brain for adaptation, bipedal mobility to be more efficient in long treks, and an increase of meat and fat in their diet for more energy as evident from their shorter gut.

Another way to put it is that at an early stage, we became generalists coping with a parade of several niches that we had to adapt to. As a result, the forest people stayed in the forest, while others roamed the deserts like today's nomads in the Sahara but retained the ability to interbreed.

The Chimpanzee: Behavioral Characteristics

There are three major factors influencing our behaviors. First, there are our emotions, which were borne within us over thousands if not millions of years in response to the environments we found ourselves in and preserved in our genetics.

Second, there are the factors of physical actions taken to express our emotions also stored away in our genetics and evolved over thousands if not millions of years in our struggle to adapt to the parade of niches we found ourselves in.

Third, there is a belief system that supports the verbal or physical expression of these emotions colored by the local environmental conditions.

For example, when an individual takes action to discriminate against another, this action evolved to protect the individual and the tribe or clan from a foe. That behavior is not recognized today as legitimate but was legitimate then and is a clue to our true niche.

You see when we feel hate, love, or indifference; we must realize that these feelings evolved in us over long periods of time for the purpose of participation in nature's orchestra and performing the cyclical dance. Emotions, actions, and belief systems are nature's answer to the song that defines us and our niche.

We must peel away the layers of inhibitions cast upon us by those who blindly saw and built mega nations that invented colossal machines to rape the Earth's treasures.

Looking for the original human behaviors adapted to their niche of zero ecological footprint is much harder than looking for the physical characteristics because there is very little in the way of artifacts. What is available to us from our past are the behaviors of our closest relatives (chimpanzees), the results of our tool making, some pictures on the walls of caves, the behaviors of our cohorts hunter-gatherers, and our own "modern" human behaviors to give us some clues.

Chimpanzees

The chimpanzee is humans' closest living relative. These two species look alike in many ways, both in body and behavior. But for a clear understanding of how closely they are related, scientists compared their DNA, an essential molecule that's the instruction manual for building each species. Humans and chimps share a surprising 98.8% of their DNA. How can we be so similar and yet so different?

Humans and chimps descended from a single ancestor species that lived 6 or 7 million years ago. As humans and chimps gradually evolved from a common ancestor, their DNA, passed from generation to generation, changed too. In fact, many of these DNA changes led to differences between human and chimp appearance and behavior.

Chimpanzee groups in different areas share different cultures much like human cultures. Tool use is a good example. Chimpanzees in one area use long twigs and alter them for better termite fishing. In a different forest chimpanzees are more often seen nut-cracking with rocks on flat surfaces.

Chimpanzees develop different cultural practices depending on their environment, and transmit these practices as learned behavior. Chimpanzees have exhibited as many as 39 learned behaviors, including feeding, mating, grooming, and tool use.

But what kinds of behaviors do most sub-species and groups of chimpanzees share? Group structuring, communication, and hunting practices are often common behaviors from one chimpanzee group to another. In addition, these behaviors aren't perfectly identical and have adaptive variations.

Chimpanzee behavior is so complex because a chimp's mental capacity is so greatly developed. Through research, many mental traits that were once considered unique to humans have been demonstrated by chimpanzees, such as reasoned thought, abstraction, generalization, symbolic representation, and a concept of self.

One of the most important discoveries was that chimpanzees make and use tools. For example, chimps pick up small twigs, strip off the leaves, and use them as tools to fish for termites in the ground, which they then sweep into their mouths as a snack.

In general, chimpanzees use objects—stems, twigs, branches, leaves, and rocks—in nine different ways to accomplish tasks associated with feeding, drinking, cleaning themselves, investigating out-of-reach objects, and as weapons.

Chimps also make sponges by chewing leaves and then dipping them into puddles of water, which they can then drink. They also use sticks and rocks to smash fruits or hard shells. Tools, however, are not used only for feeding. Adult males, for example, can enhance charging displays by hurling sticks, branches, or rocks to intimidate others.

These behaviors, passed from one generation to the next through observational learning, can be regarded as examples of chimp culture. Chimpanzees are highly social beings, just like humans. Social interactions are essential in a chimpanzee's development, learning, and overall well-being.

Chimpanzees live in social groups called communities or unit troops. At one wildlife preserve, the number of individuals has ranged between 40 and 60. Communities may be smaller or larger in other areas.

Chimpanzees' social structure can be categorized as "fusion–fission." This means that they travel around in small subgroups (of up to 10 chimps) whose membership is always changing as individuals wander off on their own or join other groups. At times, many of a community's members come together in large excited gatherings, usually when fruit is available in one part of the range, or when a sexually popular female comes into a period of female sexual receptivity.

Individuals may switch groups on occasion, but close, supportive, affectionate bonds also develop between family members and other individuals within a community and can last a lifetime. Chimpanzee family bonds are strong, especially mother–daughter bonds. Mothers and their dependent young up to age seven or so are always together. Some individuals travel together more often than others, such as siblings and pairs of male friends. Contact is maintained between members of the scattered groups by means of a distance call: the pant hoot.

Females disperse from the natal group once they are mature and spend most of their time alone, with dependent offspring. Males usually remain in birth groups, cooperate in defense of the community range, and spend long periods in proximity to other males. Males will sometimes form coalitions to support one another during conflicts with other groups.

Within a chimp community, a male hierarchy, ordered more or less in linear fashion, establishes social standing, with one male at the top or "alpha" position. All adult males dominate all females although females have their own hierarchy, albeit much less straightforward.

Age is a deciding factor in male dominance hierarchies: the alpha male is usually between the age of 20 and 26. Other factors that determine dominance and social status are physical fitness, aggressiveness, fighting skills, the ability to form coalitions, intelligence and other characteristics. Status is either maintained or changed through communication and social interactions, such as physical competition and grooming.

The males of a community regularly patrol their territories, and if they encounter individuals of a neighboring community, they may attack with extreme brutality. The only individuals who can move freely between communities are adolescent females who have not yet given birth. They may transfer to a new community permanently or, having become pregnant, move back to their own birth group.

There are several mating patterns seen in chimps. Some females in the period of sexual receptivity are more attractive than others. A popular female may be accompanied by many or all the adult males of her community, with adolescents and juveniles tagging along. Or, the dominant male of the group may show possessive behavior toward her, trying to prevent other males from mating with her. A third mating pattern is when a male persuades a female to accompany him to a peripheral part of the community range. If he can keep her there until the time of ovulation, he has a good chance of siring her child. Even low-ranking males can become fathers in this way, if they have the skill to lead a female away during the fertile period of her reproductive cycle.

Given that there is a hierarchical system in chimp societies, most disputes within a community can be solved by threats rather than actual attacks. They use gestures and postures to indicate a threat, such as tipping the head, making hitting gestures, flapping hands in the air, swaying branches, throwing objects, and charging toward another. These gestures are often combined with vocalizations.

Chimpanzees, however, are capable of physical violence. In one situation, a 4-year territory war between two groups of chimps ended with one group's killing all members of the other group. This was the first recorded account of nonhuman primate warfare.

Infanticide also occurs within chimp communities. Male chimps sometimes kill infant chimpanzees, for a variety of hypothetical reasons, but this behavior is most commonly thought to induce the female who he is mating with to wean his offspring sooner and to ensure that the offspring is his.

Chimpanzees communicate with a wide range of verbal and nonverbal means, such as calls, postures, and gestures. Chimpanzees use verbal communication, such as alarm calls, mating calls, and greeting vocalizations. So far, researchers have identified more than 30 calls, which can sometimes be heard up to 2 miles away.

Food calls (a mixture of grunts, barks, and pant hoots) alert other chimpanzees to the whereabouts of food sources. A special intensity of excited calls of this type indicates that there has been a successful kill after a hunt.

A loud, long "wraaaa" call is made when a chimpanzee comes across something unusual or dangerous. When young chimpanzees play, they emit breathy laughter. And soft grunts uttered by foraging or resting chimpanzees probably serve to maintain communication within the group.

Each individual has his or her own distinctive pant hoot, so that the caller can be identified with precision. Postures, gestures, and facial expressions communicate many messages and emotions within a group.

When greeting a dominant individual after an absence or in response to an aggressive gesture, nervous subordinates may approach with submissive signals (crouching, presenting the rump, holding the hand out) accompanied by pant grunts or squeaks. In response, the dominant individual is likely to make gestures of reassurance, such as touching, kissing, or embracing the subordinate.

Friendly physical contact is crucial in maintaining good relationships among chimpanzees. For this reason, social grooming is probably the most important social behavior, serving to sustain or improve friendships within the community and to calm nervous or tense individuals. The grin of fear seen in frightened chimpanzees may be similar to the nervous smiles given by humans when tense or in stressful situations.

When angry, chimpanzees may stand upright, swagger, wave their arms, throw branches or rocks: all with bristling hair and screaming or with lips bunched in ferocious scowls. Male chimpanzees proclaim their dominance with spectacular charging displays during which they slap their hands, stamp their feet, drag branches as they run, or hurl rocks to perhaps intimidate a higher-ranking individual without having to fight.

Chimps are omnivores and eat not only fruits, nuts, seeds, blossoms, and leaves but many kinds of insects and occasionally the meat of medium-sized mammals. Chimpanzees, like humans, have such diverse tastes that they are able to live in a wide variety of habitats, unlike gorillas and orangutans, which have more specialized diets in the wild.

Male chimpanzees occasionally hunt cooperatively, to kill and eat other mammals, such as bush pigs, monkeys, and antelope. For instance, a group of chimpanzees attacked, killed, and ate a red colobus monkey that had climbed high into a tree. The hunters covered all available escape routes while one adolescent male crept up after the prey and captured it, whereupon the other males instantly rushed up and seized parts of the carcass.

Successful hunters typically share some portion of their kill with other group members in response to a variety of begging behaviors. Most of the captured animal is eaten, including the brain. Meat is a favored food item among chimpanzees, but makes up less than 2% of their overall diet.

An interesting observation has also been made where chimps deliberately eat medicinal plants like *Aspilia* leaves and minerals to relieve stomach pains or reduce the number of internal parasites.

Chimps are also capable of walking upright on two legs, like humans. Unlike humans, they usually use this mode of locomotion only if they need to travel while carrying objects in their hands.

Hunter-Gatherers

Over 3 million years ago, our ancestors began using tools made of stone and bone that seem to be associated with hunting and digging. This early hunting and gathering played a big part in humans' evolving from *H. erectus*, who had spread throughout Africa and then Europe a million years ago, to eventually become *Homo sapiens*, who subsequently spread throughout the world. Beginning around 10,000 years ago, humans began domestication of animals and plants, becoming farmers and herders. The human lifestyle changed dramatically, with towns and cities and empires developing quickly. By 1500 A.D., at the dawn of European expansion, hunter-gatherers still occupied almost one-third of the world's landmass, including all Australia, the northwestern half of North America, and the southern part of South America, including parts of sub-Saharan Africa and parts of Asia. Over time, hunter-gatherers were able to survive with their cultures intact only in areas that were of little interest to the rest of humanity, such as places with little water, thick jungles, impassible mountains, and frozen Arctic areas. In addition, attempts to study these so-called primitive tribes have been lethal, where up to 50% of their populations die of so-called modern ordinary diseases such as the flu.

The Organization of Hunter-Gatherer Groups

The hunter-gatherer lifestyle lasted for 3 million years and proved itself to be at zero ecological footprint status by surviving that way for so long without damaging the environments. Some hunter-gatherers still exist today hiding in various inaccessible geographic regions. These hunter-gatherer societies are typically organized by having small bands of ~30 individuals that are quite independent, but connected to a larger tribe of several or many similar groups.

Individuals may move from one band to another, and often young adults will select their spouses from other bands within the larger tribe. These bands will usually move from place to place in a seminomadic way although some may remain in permanent settlements. Each band has a leader, or chief, who usually is an organizer and mediator, but has little power beyond that. These groups hardly possess any stockpiles of material wealth, and leaders particularly avoid having any more possessions than other members of the group do, so that there will not be any resentment or jealousy in the group. Accordingly, a chief usually does not have much power over the rest of the members of the group. All members of the band have an equal voice in decision making. The chief only organizes those voices. In fact, if one person becomes too aggravating to another person of the tribe, that person can simply move to another band. The chief is always under pressure to be humble and accommodating because he does not want migration of people away from his band.

The greater complexity of food-producing groups as compared to those who hunt and gather is evident in their types of leadership. Hunter-gatherers typically live in bands without a central leader. Instead of a ruler, leadership is in the form of respected individuals who can persuade the group to do things, but who have no real power. Given that everyone has a relatively equal share of the available resources, hunter-gatherers are egalitarian. In contrast, horticulturalists and pastoralists generally live in tribes, which are larger, more organized groups either without a central leader or in chiefdoms in which one individual has absolute power that is inherited. There must be a surplus of food—generally obtainable only through agriculture—for a chiefdom to exist because the chief increases his control by collecting the surplus and redistributing it. The hierarchy of power means that there is less equality in food-producing societies (as opposed to hunter-gatherers), so that resources are not always evenly shared. Agriculture provides the circumstances where disparities of wealth among households are possible, and the wealthier households begin to exploit poorer units.

Cave Paintings

These are paintings found on cave walls and ceilings, and especially those of prehistoric origin, which date back to some 40,000 years ago in both Asia and Europe. The exact purpose of the Paleolithic cave paintings is not known. Evidence suggests that they were not merely decorations of living areas because the caves in which they have been found do not have signs of ongoing habitation. They are also often located in areas of caves that are not easily accessible. Some theories hold that cave paintings may have been a way of communicating with others, whereas other theories ascribe a religious or ceremonial purpose to them. The paintings are remarkably similar around the world, with animals being common subjects that yield the most impressive images. Humans mainly appear as images of hands, mostly hand stencils made by blowing pigment on a hand held to the wall.

The most common subjects in cave paintings are large wild animals, such as bison, horses, aurochs, and deer, and tracings of human hands as well as abstract patterns, called finger flutings. The species found most often were suitable for hunting by humans, but were not necessarily the actual typical prey found in the associated deposits of bones; for example, the painters of a site in France have mainly left reindeer bones, but this species does not appear at all in the cave paintings, where equine species are the most common. Drawings of humans were rare and are usually schematic as opposed to the more detailed and naturalistic images of animal subjects. One explanation for this pattern may be that realistically painting the human form was forbidden by a powerful religious taboo. Other geologists suggest that climate controlled the themes depicted. Pigments used include red and yellow ochre, hematite, manganese oxide, and charcoal. Sometimes, the silhouette of an animal was incised in the rock first and in some caves all or many of the images are only engraved in this fashion.

Similarly, large animals are the most common subjects in the many small carved and engraved bone or ivory and stone pieces dating from the same periods. But these include the group of Venus figurines, which have no real equivalent in cave paintings.

Hand stencils, made by placing a hand on the wall and blowing pigment at it, form a characteristic image of a roughly round area of solid pigment with the uncolored shape of the hand in the center, which may then be decorated with lines or dashes. These are often found in the same caves as other paintings, or may be the only form of painting at a location. Some walls contain many hand stencils. Similar hands are also painted in the usual fashion. A number of hands show a finger wholly or partly missing. Some interpreted the paintings as being hunting magic, meant to increase the number of animals.

Another theory, developed on the basis of ethnographic studies of contemporary hunter-gatherer societies, is that the paintings were made by Paleolithic shamans. A shaman would retreat into the darkness of the caves, enter a trance state, and then paint images of his or her visions, perhaps with some notion of drawing power out of the cave walls themselves.

Studies of both highly artistic and publicized paintings and a variety of lower-quality art and figurines identify a wide range of skill and ages among the artists. There is a hypothesis that the main themes in the paintings (and other artifacts) such as powerful beasts, risky hunting scenes, and the representation of women in the Venus figurines are the work of adolescent males, who constituted a large part of the human population at the time. Nonetheless, after analysis of hand prints and stencils in French and Spanish caves, it has been proposed that a proportion of them are female hands.

Summary

Our original zero ecological footprint niche was apparently one of hunter-gatherer nomadic small groups consisting of several families that foraged on fruits, berries, mushrooms, some grains, flowers, and game. Hunter-gatherers move on when these food sources reached a stage of reduction but were still able to recover.

These small groups had territories that were defended fiercely because the territories were needed for survival as small groups and did not represent an object of wealth or status. The territory protected their most simple of natural cyclical lifestyles.

Tools were made of local perishable materials consisting of tree branches, roots, animal skins, animal bones, and stones. The process of making these tools also involved perishable and local materials to guarantee their return to nature. The same goes for the construction of dwellings and shelters.

They had a symbolic language and therefore thought, as demonstrated by the art found in caves and by contemporary hunter-gatherer groups. Today's hunter-gatherer groups are healthy and free of modern human diseases such as obesity, hypertension, diabetes, and communicable diseases.

They had music and song in their lives that told stories of the universe, the hunt as well as those of love and appreciation. They had rituals for preserving the group and the environment. You may think of this as their religious demonstration.

The context of the hunter-gatherer lifestyle within its environment is the one that best describes the natural human niche that is in balance with nature supporting the zero ecological footprint way of life through the cyclical character of life on Earth.

Return

Now that the music for our dancing cycles has stopped and was replaced by the clanking noise of gargantuan machines, it will take some effort to return to how it once was. It is also said that "You can never go home again." The best we can hope for is a new way of life perhaps a little different from what it was 40,000 years ago and certainly radically different from what it is today.

How we get there from here will be discussed in the next chapter. For now we need to cover the rules under which we need to live and thrive followed by the benefits reaped thereof.

The first thing is our attitude. We must become wise to the fact and consider ourselves a part and not apart from the living cycles, the web, and the biota on Earth. Notions like "humans are at the top of the food chain" and "humans have superior intelligence" have continuously crept into everyday talk and the literature as an expression of our naturally blind drive to acquire more.

It must be emphasized that humans are not at the top of the food chain but they are a link no better or worse than any other link, be it a clump of mud or another species in the web of life. As for human intelligence, it has been subservient to the opportunistic drive borne in all species, be they amoebas, birds, or dinosaurs. Now is the time to forgo this opportunistic lifestyle and develop a mature attitude.

Second, each one of us must identify their living cycle. It is the living cycle of the omnivore or those who consume plant and animal matter for sustenance and leave behind only the waste of plant and animal matter for our brethren who use it for their sustenance.

Third, we must not ingest or consume anything that did not originate from our local living cycle. That means we must not ingest anything processed, manufactured, or imported because this kind of nutrition only returns us into the endless loop of self-destruction. We must thank our sources of food, Earth, and the sun because they are the providers.

Fourth, we must not leave behind anything that can't be consumed by those who are next in line in our local living cycle. It shouldn't be processed or manufactured.

Fifth, we must not use implements that do not come from our local living cycle. We can use anything that occurs naturally such as local stones, tree branches, leaves, and the like.

Cyclical Lifestyle Benefits:

The following are the benefits gained by the human condition following the five rules stated above. They don't cover the obvious benefits of the planet Earth and the rest of its biota.

- We will hunt and gather around where we live and so we do not have to leave home base or be away from family and friends.
- Hunting and gathering takes about 2 to 4 hours a day and all the food is fresh. With a short work day, we have abundant leisure time to be with family and friends.
- Food is always fresh and better than our "Organic"-labeled foods: without hormones, pesticides, or fertilizers.
- The elderly spend their days not alone but as a part of the group they have always lived with.
- Hunters and gatherers are always physically fit, and there is no problem of excess body weight.
- Because hunter-gatherers require large areas as their territories, we won't suffer from the sedentary-lifestyle diseases of civilization.
- Because we won't have any possessions, we will have no worries about anyone stealing.
- Men get to exercise their instincts for hunting and fishing.
- Men and women get to exercise all their instincts for living.
- Get rid of the stress of modern life.
- Forgo death by technology.
- Forgo living in an untested, ever changing, and artificial world.
- Enjoy personal satisfaction that each one of us will be living an Earth-sustainable lifestyle that will last for thousands if not millions of generations to come.

- Enjoy personal satisfaction that each one of us will be living a lifestyle that is a gift of life to thousands if not millions of generations to come.
- Experience personal satisfaction that each one of us will have available to them wholesome and healthy foods whose sources will last for thousands if not millions of generations to come.
- Live a more energy-efficient lifestyle.

Only when we live this simple life, will we sway to the music and join the rest of the biota in the dance once again. So, praise the sun and mother Earth because we are their precious children one and all.

Path

It would help to define what the population size target is for this chapter. We have stated in a previous chapter that the most energy-efficient and zero ecological footprint population size needs to be from 10 to 50 million human individuals scattered across the Earth adopting the hunter-gatherer lifestyle or something similar. Given that the current population is approximately 7.5 billion, the task in front of us is extremely daunting to say the least.

We have two major options in front of us at this time. The first is to do nothing. There is evidence that nature takes care of its own problems, and therefore this approach might be a viable way to go and we will explore this option presently from a logical point of view using the data we have at hand. The second option of doing something that will be covered later.

The United Nations did a simulation of birth and death rates in the human population, which indicated that the human population bloom will peak at 10 billion, at which point births will equal deaths. That is only ~35% more than what the human population count is right now. Some may think that this is a small increase that may be supported. At that point, however, humanity will be consuming twice what the Earth can produce. This means that the rate of consuming the infrastructure of Earth will double.

Other simulations indicate that this situation will lead to degraded living conditions. For example, health care and basic food needs will not be accessible to a growing number of humans, the death rate will continue to grow, and the population will decrease until it gradually reaches ~2 billion humans in approximately 50 years.

Along the way from a population level of 7.5 to 10 then down to 2 billion humans, resources will be scarce; the probability of social unrest will increase, leading to degradation of the social fabric, an increase in anarchy, and the specter of nuclear wars. These scenarios will lead to an uninhabitable Earth in the form of a decimated nonhuman species and a poisoned physical environment as discussed earlier.

The second option is to do something to avert the destruction of our Earth and its inhabitants as they stand today. To be effective we need to include all humans in the solution we need to educate everyone. All people need to know the tragic consequences of doing nothing. They also need to know how they can help to find a solution to (and prevent) the upcoming disaster.

In the face of the upcoming shortages in economically extractable metals and minerals from the Earth, it has been suggested that humanity mine these resourced from other planets or asteroids. Unfortunately, this approach adds to the energy consumption of each human and thereby decreases their efficiency and goes against the Earthly bio-fuel-efficiency trend.

Some advocate sending a group of astronauts to a distant planet to start a human colony. That idea is natural. It represents our opportunistic spirit much like the slime mold's fruiting body using the wind or other animals to carry its spores to distant territories richer in resources. Unfortunately, planets that are assumed to support Earthlike living conditions are light years away, and it will take more than 100 years of research to make such a journey possible, meanwhile this effort will be interrupted by famine and unrest hampering any hopes of funding such an expensive endeavor.

As discussed, the use of fossil fuels to run our industrial economies cannot come about without the support of the technology creation sector. They figured out how to use energy to make an aspirin tablet to sending men to the moon on a fuel-guzzling rocket. They increased the fossil fuel consumption from zero to 14 billion calories per person per year _making humans by far the least efficient consumer of energy of all the living on planet Earth_. They should have the opportunity to redeem themselves and create technologies that will reduce fossil fuel consumption back to zero again.

And indeed, they have been trying. There has been a lot of talk of a carbon tax that imposes a monetary penalty on carbon emissions from burning of fossil fuels. The way it works is that a government agency taxes fossil fuel companies (that pump this fuel or dig it out of the ground). The amount of tax is dependent on the amount of carbon emissions the fuel will eventually produce when used. That agency then pays companies for the "scrubbing" the carbon before or after it escapes into the air.

Unfortunately, the carbon tax does not decrease the production of fossil fuels, and in some cases, increases it. This situation adds to the human bloom and its side effects of consuming the infrastructure of the Earth and decreases the overall efficiency of humans.

What is needed is a method of gradually <u>reducing</u> fossil fuel use to zero. One way to do that is to reduce carbon emissions by 70 pounds per person per year every year for 30 years. This approach will result in two achievements. The first is that in the year 2031, there will be zero fossil fuel use. The second is that possibly the planet will be saved and able to recover (see Appendix 4).

Possible Outcome if We Aim for Zero Fossil Fuel Use in 30 years

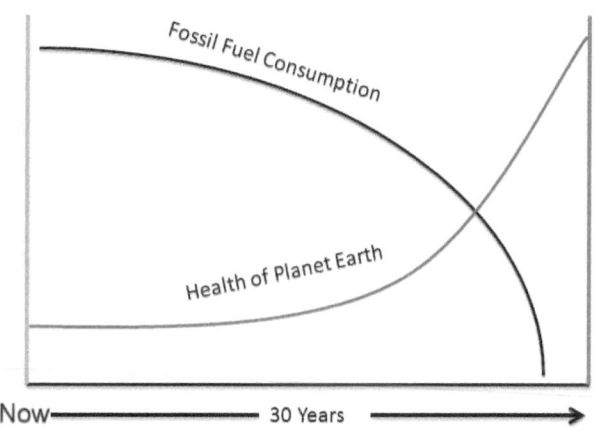

Now ——————— 30 Years ——————⟶

Another opportunity is the altruistic act of voluntary zero births through mass contraception except for a select group of humans who will act as our seed for generations to come much like the slime mold's fruiting body. The difficult task here is the selection process. It is our hope that all so called "primitive" tribes be exempt from this effort. They have survived at zero ecological footprint the longest, are best equipped physically, socially and have the right belief systems to be the most suitable seed.

No single approach from the opportunities presented above will work by itself. The best opportunities are those of education, voluntary birth control and the efforts to eliminate fossil fuel use are the best. The path to these goals is difficult to say the least causing social disruption and mass population relocation since the present world economies are largely dependent on these fuels. But we have a duty to insure a favorable outcome for the Earth's biota including our own sooner rather than later.

Proposals

Education

Education is the oracle for the future. Without it we are sailing on a ship with no compass. This education must be one of integrity and not based on alternative facts. It must be rest upon the truth and not political aspirations. It must emanate from those who dedicate their lives in the pursuit of the scientific method and not the consensus of the masses.

It is true the scientific community is struggling in the wake of the revival of faith based political activists who want to protect the status quoi and change things to the way they were in the nostalgic past. Unfortunately that is like having the kidneys dictating the judgements of the brain. What should be happening is that decisions be made based on our scientifically evaluated information. That is the way nature intended it to be.

All species on Earth have an information system based on knowledge. Single celled species have a genetically based information system that was honed by experience over millions of years. More complex species require a central nervous system for gathering information and making decisions in a shorter time frame. The more complex species, like humans, have a brain that receives external and internal evidence to make timely choices on how they should respond; the more accurate the incoming data the better the resultant actions.

To the human species, the scientific method is currently the only approach to understand the truth of nature as opposed to consensus of the masses that only yields mediocrity. Governments should consist of meek men of science, 45 years old serving two years each as conscripts with a modest salary. Only then can the human species move in the required directions of zero ecological footprint, restoring the cyclical nature of life on Earth and become efficient energy users once again. For example, in the United States of America, the constitution should be changed so that all House Representatives, Senators and the President should be replaced by such scientists.

The tasks before these scientist rulers are to educate the public in the upcoming dark 30 years, prepare them for almost zero births and zero fossil fuel use. Of course the reason why all these changes need to occur has to be loud and clear. The public must be involved and engaged on a global scale.

Almost Zero Human Population Birth Rate Plan

The aim is to reduce the birth rate to zero for those populations that used fossil fuels. It is proposed that every birth is to be taxed at one year's salary of the father.

Zero Fossil Fuel Use Plan

The aim is to reduce fossil fuel use to zero in 30 years. It is proposed that each consecutive fossil fuel production year be reduced by 0.239 times the previous year's total production. This will achieve zero fossil fuel production in 30 years. Any fossil fuel company producing an excess of this reduced amount will be taxed 0.239 of its total production for that year.

The above are my proposals. What are yours?

Now is the time we should start being efficient catalysts and rejoin the rest of the biota claiming our equitable allotment of the Sun's blessing. Now is the time we start achieving a zero ecological footprint and nestle back in with our brothers and sisters for a long term future. Now is the time to wholeheartedly rejoin the symphony of life on Earth and dance in unbroken cycles.

Appendix 1

Birth of Economic Thought

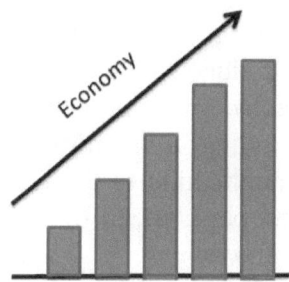

According to Hamid S. Hosseini, the power of supply and demand was understood to some extent by several early Muslim scholars, such as 14th-century A.D. Mamluk scholar Ibn Taymiyyah, who wrote: "If desire for goods increases while its availability decreases, its price rises. On the other hand, if availability of the good increases and the desire for it decreases, the price comes down."

John Locke's 1691 work, *Some Considerations on the Consequences of the Lowering of Interest and the Raising of the Value of Money* includes an early and clear description of supply and demand and their relation. In this description, demand is rent where the price of any commodity rises or falls by the proportion of the number of buyers and sellers and that which regulates the price of goods is nothing else but their quantity in proportion to their rent.

The phrase "supply and demand" was first used by James Denham-Stuart in his Inquiry into the *Principles of Political Economy*, published in 1767. Adam Smith used the phrase in his 1776 book, *The Wealth of Nations*, and David Ricardo titled one chapter of his 1817 work, *Principles of Political Economy and Taxation*, "On the Influence of Demand and Supply on Price."

In *The Wealth of Nations*, Smith generally assumed that the supply price was fixed but that its value would decrease as its scarcity increased, in effect what was later called the law of demand. Ricardo, in *Principles of Political Economy and Taxation*, more rigorously laid down the idea of the assumptions that were used to build his ideas of supply and demand. Antoine Augustin Cournot first developed a mathematical model of supply and demand in his 1838 research into the *Mathematical Principles of Wealth*.

In his 1870 essay, *On the Graphical Representation of Supply and Demand*, Fleming Jenkin introduced a diagrammatic method into the English economic literature by publishing the first drawing of supply and demand curves, including comparative statistics from a shift of supply or demand and application to the labor market. The model was further developed and popularized by Alfred Marshall in the 1890 textbook *Principles of Economics*.

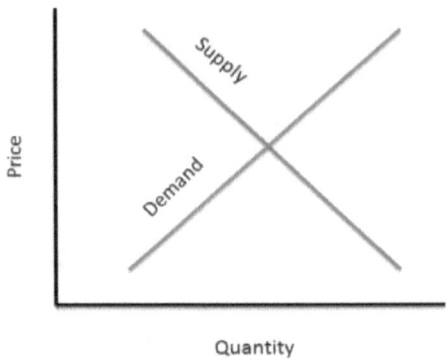

The Supply and Demand Chart

Within the context of supply and demand is the need to increase the bottom line of those supplying the products and in turn the consumers who are demanding cheaper units. To the manufacturer or producer, this means a decrease in the cost of producing a unit. To the consumer, it means a decrease in the cost of purchasing a unit.

Appendix 2

Energy

The Earth is a predominantly closed system with energy coming in from outside. All other resources on Earth, therefore, need to be reused endlessly. If the Earth had a prime directive, it would be to use it, reuse it, and recycle it.

Life is energy, and life on Earth comes mostly from the energy the sun throws its way. Through photosynthesis, plants capture this energy and thus start its journey of flowing through us all.

Matter is in various states of potential energy. The word "potential" means that an atom or molecule can give up energy under the right circumstances. To visualize this situation, you can think of a mountain with boulders lying in gullies on its sloping sides. Boulders at the top of the mountain have the potential of rolling down longer distances and with more energy than others that are, for example, half way up.

Some gullies are deep and hold the boulders more securely in place than others. When a boulder rolls down the mountain, it moves from a higher state of energy to a lower one. One way this transition can happen is by applying some kinetic energy to the boulder by digging it out of its gully with a shovel. Another way this can happen is if it rains and a side of the gully dissolves allowing the bolder to break free and roll downhill.

In biological systems, enzymes and other compounds exemplified by the shovels above release energy from energy-rich molecules to be used in metabolism and catabolism. The energy released by each reaction is extremely small and under the control of the living cell. In biological systems, we talk about grains of sand instead of boulders on the side of a mountain.

So, how do the grains of sand get up the mountain in the first place? With the help of energy packets called photons that come from the sun and specialized cells found in plant leaves; carbon dioxide and water are converted to sugars, which are chemical compounds with higher potential energy and, therefore, higher up the mountain.

The simplest of sugars is glucose. It is a ring with six sides of mostly carbon with hydrogen and oxygen hanging off the vertices. Two other compounds that are higher up the mountain originate from this simple sugar. They are starches and cellulose.

A schematic of the composition and relations of atoms in a sugar molecule

Starches are long chains of glucose rings used to store energy in the seeds and other organs such as roots.

Cellulose has a more complicated structure than do starches originating from two-bonded long chains of glucose rings. Cellulose provides support structures such as stalks, branches, and trunks for plants.

Another compound is a fatty acid. Fatty acids are simple carbon and hydrogen strings that end in a hydroxyl group. Fatty acids are components of fat. Fat is another compound for storing energy. A type of fat consists of three long strings of fatty acids held together at one end by a group of atoms called the glycerol group.

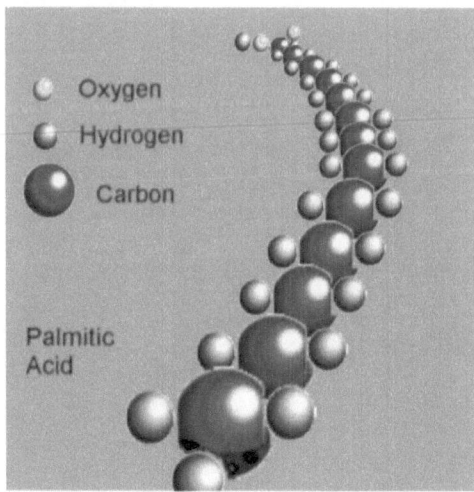

A schematic of the composition and relation of atoms in fatty acid molecules

A schematic of the composition and relation of atoms in a molecule of palmitic acid

Photosynthesis

Photosynthesis is the process by which plants, some bacteria, and a few other organisms use the energy from the sun to produce sugar used by plants and most other living organisms. A green pigment called chlorophyll creates the sugar. Most of the time, the photosynthetic process uses water and releases oxygen. More precisely, the conversion of six molecules of water plus six molecules of carbon dioxide by photosynthesis produces one molecule of sugar plus six molecules of oxygen.

Respiration

Respiration is the metabolic (building) reaction and process that builds tissues and maintains cell functions that take place in a cell and obtains the energy required from sugars. Energy is released by the oxidation of fuel molecules such as sugars and is stored in energy carriers such as adenosine triphosphate (ATP) that are then used for many processes requiring energy, including biosynthesis, locomotion, or transport of molecules across cell membranes.

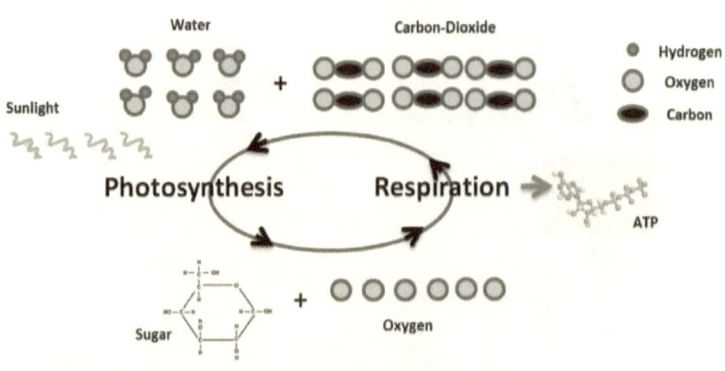

The Photosynthesis-Respiration cycle

In summary, respiration takes a sugar molecule and six oxygen molecules and produces six carbon dioxide, six water, and a bunch of ATP molecules.

So far, we have described how the living, with the help of the sun, push matter up the mountain to store energy. The path up is not exactly the same path going down where the living start using this stored energy and releasing the carbon, hydrogen, and oxygen atoms captured in photosynthesis and become part of natural cycles. A cycle is a term used to describe the life of a compound from birth to death and then rebirth in a way that results in no net accumulation or loss over the long run.

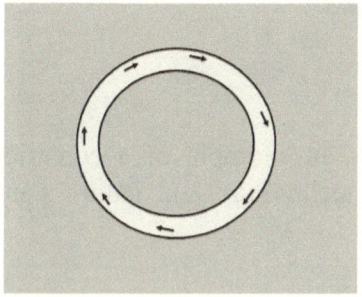

The above schematic is an example of a free flowing natural cycle

Technological advances and sudden high rates of resource consumption made by the human population explosion have caused a net increase of compounds related to the natural cycles to be produced that have far outpaced Earth's ability to recycle them in the way it has for billions of years. In addition, many of the new technologies produce chemicals that have no known cycles and thus they are accumulating and often resulting in dangers to life on the planet.

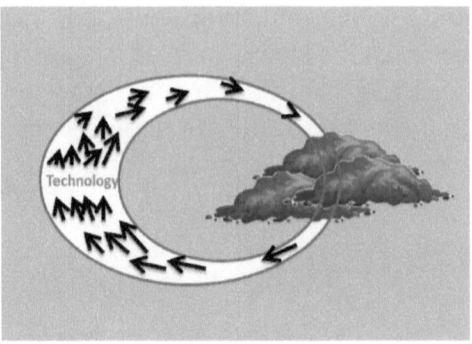

The above schematic is an example of a constricted cycle caused by man's appetite for technology and nature's inability to process its byproducts.

Appendix 3

The Target Human Population Count

How can humans achieve zero ecological footprint? We will try to do that via two approaches. One approach is to take the sun's energy allotment for humans and calculate the number of humans who should be living within the Earth's boundaries. We will call this the top-down method.

The second approach is to take a stable long-lasting human population that exists today and extrapulate from it the number of humans that can be supported on Earth. We will call this the bottom-up method.

Top-Down Analysis

Rank	Category	# Individuals (Billions)
1	Blue whales	0
2	Cattle	1
3	Sheep & Goats	2
4	Humans	7
5	Chickens	24
6	Ants	10,000,000
7	Antartic crill	800,000,000,000,000
8	Bacteria	5,000,000,000,000,000,000,000,000,000

A table of Earth life forms ranked by individual counts

Rank	Category	Biomass (Wet million tones)
1	Blue whales	36
2	Chickens	48
3	Sheep & Goats	105
4	Humans	350
5	Antartic crill	379
6	Cattle	520
7	Picoplankton	1,000
8	Ants	5,000
9	Bacteria	680,000

A table of Earth life forms ranked by biomass

To reduce the ecological footprint of humans to zero, we need to talk about the human population, and its allotment of the energy from the Sun.

It must be realized that the rays of the sun are what makes the whole Earth come alive and the amount of sunlight coming to Earth is approximately 10,547,945,205,479,500,000,000 joules/day where only 71% is absorbed by the planet which is equal to 7,489,041,095,890,410,000,000 joules/day. Please note this figure represents both land-based and ocean-based rays. (1)

Plants are converters of this energy into plant matter and their average photosynthetic efficiency is approximately 4.5%. That leaves us with 337,006,849,315,069,000,000 joules/day for all the living on Earth (2).

The biomass of present humans on Earth is approximately 0.0509396% of all the living creatures. This figure represents approximately 7.4 billion individuals which many experts consider too many to be sustainable. I will use this percentage as a starting point and consider it in the sum of all human biomass plus the biomass of all its food resources (3).

Therefore, the basic quota of energy for consumption by humans is 171,670,000,437,024,000 joules/day (4).

Given that

A) The percentage of animal matter in the human diet is 5.0%

B) The percentage of plant matter in the human diet is 95.0%

C) There are 4.184 joules in a calorie

D) Calorie intake of a human is approximately 1,800,000/day (The calories count you and I used in our daily lives is 1,800/day)

E) Humans consume 5% of growing plant matter (ratio of the whole plant to the grain is 20:1)

F) The ratio of plant matter required to produce edible animal matter is 2.5% (ratio of plants to edible meat is 40:1)

Energy consumption per human is at 753,120,000 joules/day (5).

Given that the human energy consumption efficiency is 20% (6), the total number of humans that can be supported based on their proportion of the total Earth biota is approximately 45,589,016 individuals (7).

The total land mass of arable land that supports the human population is 15 million kilometers squared. Therefore, each human on average occupies and gains resources from approximately 330,000 meters squared of arable land. (8)

Bottom-Up Analysis

The BaMbuti people live in the Ituri Forest, a tropical rainforest covering approximately 63,000 km^2 of the north/northeast section of the Democratic Republic of the Congo, Africa. BaMbuti are pygmy hunter-gatherers, and are some of the oldest indigenous people of the Congo region. They are composed of bands ranging from 15 to 60 people, with a population hovering around 30,000 to 40,000. It is not known how long they, the pigmy people, have been living in the Ituri forest, but there have been reports of their existence since the pharaohs of Egypt approximately 4,000 years ago (*The Forest People* by Turnbull, Colin M.).

Deliverance

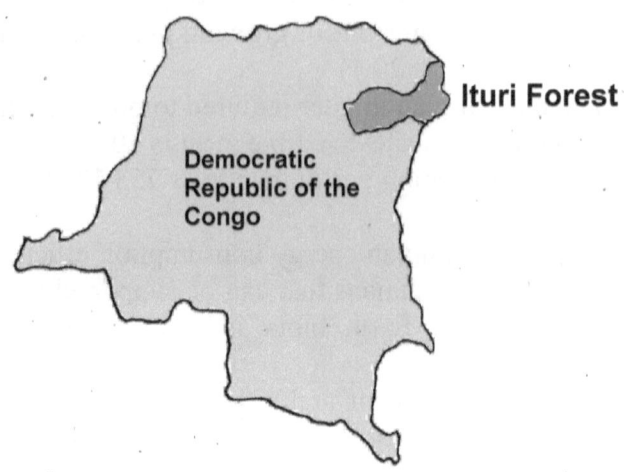

Location of the Ituri Forest in the Democratic Republic of the Congo

The Ituri Forest receives a large amount of rainfall annually, ranging from 50 to 70 inches. The dry season is relatively short, ranging from 1 to 2 months. The forest is a moist, humid region strewn with rivers and lakes. Too much rainfall, as well as droughts, can greatly diminish the food supply.

In the rainy season, pigmy people forage in a part of the forest, and in the dry season, they expand their foraging territories. Their wild foodstuffs include crabs, shellfish, ants, larvae, snails, pigs, antelope, monkeys, and fish as well as honey. The plant component of their diet includes wild yams, berries, fruits, roots, leaves and cola nuts. They also eat mushrooms.

The habitat area for each human in the Ituri forest is approximately 1,500,000 meters squared. Because they have been living in the forest for thousands of years with zero ecological footprint, we can try to extrapolate the population size if the whole Earth lived as they do; the population of the Earth should be approximately 10 million individuals (9).

As you recall from the top-down method discussed previously, the assumed population size was approximately 46 million individuals. It is argued here that these figures are close to one another and may represent the range of the Earth's population that can exist at zero ecological footprint.

Appendix 4

Metrics

There are different ways to express fossil fuel use. For example, we can express it in the quantity extracted from earth in tons. Unfortunately, a ton of crude oil has energy content different from that of a ton of coal. The convention, however, is to express it in terms of the carbon content of the end product namely carbon dioxide because they all end up producing that gas whether it is from crude oil or coal. The following is a table of conversion factors:

A ton of CO_2 = 2,205 pounds of CO_2
Pound of CO_2 = $2.4 * 10^7$ joules
A calorie = 4.184 joules

2004 World oil consumption was 3,900 million tones
2003 World coal consumption was 3,700 million tons of oil equivalents
2003 World natural gas consumption was 2,300 million tons of oil equivalents

On average each human consumes 64 million calories of energy per day. 41 million calories per day are from the burning of fossil fuels and 23 million calories per day are provided by the Sun.

Appendix 5

Terminology

Micro - Mezzo - Macro - Mega

This publication uses a different terminology regarding the complexity of a living organism. A species that consists of one celled individuals is designated as a "Micro" catalyst. In turn, a species that consists of two or more cells is called a "Mezzo" catalyst and a city or country that consists of multiple mezzo individuals is called "Macro" catalyst. A large living system like an ecosystem is designated a "Mega" catalyst.

Catalysts

It is explained in the chapter on Abiogenesis that the origin of life depends on catalysts and their activities. When we talk about organisms we carry through the concept of the catalyst whether it is a single enzyme, a one celled or multi celled organism or a collection of different species. They are all catalysts in one way or another.

Mature Species

A mature species is one that adapted to its surroundings in a way that does not degrade the environment making sure that natural resources are preserved for generations to come.

Appendix 6

Evolution

How is evolution achieved? How does lowering the amount of energy in a chemical reaction attained? The answer is in the availability of a catalyst. In most cases where a catalyst is required, reactions occur faster because they require less activation energy. The following six parts explain how this occurs.

Part one: The application of energy

Chemical reactions are happening naturally all the time. For example, as the temperatures increases in any particular environment chemical reactions occur more frequently. If chemicals require 10 units of energy (this is called the activation energy) to react and the available energy in the environment is 8 units the chemical reaction will not occur. If the energy available in the environment increases to 11 units then the chemicals will react.

Part two: The Catalysts

Catalysis is the increase in the rate of a chemical reaction due to the participation of an additional substance called a catalyst. In most cases where a catalyst is required, reactions occur faster because they require less activation energy. In the example above, if the activation energy for a chemical reaction is 10 units and the available energy in the environment is 8 units then the reaction will not occur. In the event that a catalyst is present in this environment which lowers the activation energy for the reaction to 7 units the reaction will proceed. Furthermore, catalysts are not consumed in catalyzed reactions therefore they can continue to act repeatedly. Often only tiny amounts are required in principle.

Part three: The Enzymes

Enzymes are a subset of catalysts. They accelerate, or catalyze, bio-chemical reactions by lowering the activation energy required for the reaction. The molecules at the beginning of the process upon which enzymes act are called substrates that are converted into different molecules, called products. Almost all metabolic processes in a living cell need enzymes.

Part four: Efficiency

It has been demonstrated that catalysts lower the activation energy of a reaction and we can say that a chemical reaction that takes less energy to convert substrate to product is also more efficient in the use of energy.

Part five: Effects of Efficiency

There is a relationship between catalysts and evolution. Say you have two catalysts in an environment. Call them C1 and C2 that catalyze products C1 and C2 respectively. C1 lowers the activation energy by 2 units causing reactions that produce 10 C1s while C2 lowers the activation energy by 4 units resulting in reactions that produce 1000 C2s. After 4 replications C2 will outnumber C1 by a factor of 16.

Part six: Self Replication

Furthermore, catalyst C2 was the daughter of C1 which was changed to be more efficient by an environmental agent.

To summarize, evolution is a process of self-replication of catalysts where a change or changes occur to the offspring that lead to more efficient catalysis.

Reading List

This reading list covers the major topics in the current and future events in human history.

2052: A Global Forecast for the Next Forty Years Paperback; Jorgen Randers; June 13, 2012

Beyond the Inflection Point: An Economic Defense of the Limits-To-Growth Theory; Andrew Currie

Cannibal Business: The Limits to Growth; Bruce Ferguson

Feeding the Fire: The Lost History and Uncertain Future of Mankind's Energy Addiction; Mark Eberhart; May 8, 2007

Limits to Growth: The 30-Year Update; Donella H. Meadows and Jorgen Randers; Jun 1, 2004

Scatter, Adapt, and Remember: How Humans Will Survive a Mass Extinction; Annalee Newitz

The End of Growth: Adapting to Our New Economic Reality; August 9, 2011

The No-Growth Imperative: Creating Sustainable Communities under Ecological Limits to Growth; Gabor Zovanyi

The Sixth Extinction: An Unnatural History; Elizabeth Kolbert; January 6, 2015

Under a Green Sky: Global Warming, the Mass Extinctions of the Past, and What They Can Tell Us About Our Future; Peter D. Ward; March 25, 2008

www.ingramcontent.com/pod-product-compliance
Lightning Source LLC
Chambersburg PA
CBHW022002170526
45157CB00003B/1112